FarmPlate

- - - - - - - - - - - - - -

VERMONT

- - - - - - - - - - - - - -

BEER

- - - - - - - - - - - - - -

BEHIND THE SCENES

with Vermont's Craft Brewers

FarmPlate

VERMONT
BEER

- - - - - - - - - - - - - - - -

Published by FarmPlate Books
www.FarmPlate.com

- -

Copyright © 2014 by FarmPlate Books

- -

A FarmPlate Guide

- -

Edited by Kim Werner
Design by Steve DeCusatis Design
Layout by Sarah Hebbel-Stone
Cover image © Bob Montgomery Images/Hill Farmstead Brewery

- -

Library of Congress Control Number: 2013954928
ISBN 978-0-9860667-0-2

Printed and bound in Canada
10 9 8 7 6 5 4 3 2 1

FarmPlate
Find Better Food Every Day

Photo credits: Aaron Rohde 42; Beana Bern Photography 90; Ben Hudson Photography 142, 144; Ben Sarle 6, 62, 72, 108, 146, 262, 264, 266; Bob Jenks (Jenks Studio of Photography) 78, 130, 132; Bob M. Montgomery Images/Hill Farmstead Brewery cover, 68, 70, 274; Brandon Fenner 50; BrewBokeh.com (Michael Donk) 8, 10, 80, 210; Brian Coon 114; Brian Eckert 56; Caleb Kenna 102; Derek Hallquist 134; Drew Vetere 84, 112; Elizabeth Campbell 138, 140; Eric Bruch Photography 28, 30; Ian MacLellan 52, 54; Jason Griffith 88; Jeff Giknis 48; Jordan Hayman 22; Lindzer Studios 44, 46; Mark Linton 76; Mark Magiera 32; Matthew Scher (Fuj On Tap) 34; Melissa Hamilton 120; Monica Donovan 82; Nicole Sauer 92, 94; Paul Varricchione 58; Sarah Hebbel-Stone 20, 40, 60, 100, 104, 270, 272; Sean Reen 124; Shawn Rice 96; Steve Hill 98; Todd Cummings 118; Tom McNeill 74; Wendy Reese 86

This book is dedicated to the brewers whose stories are told on these pages.

Their beer and their passion will inspire a new generation of aspiring brewers and discerning beer drinkers.

Thank you for being willing to share so much of yourselves.

KIM WERNER
FarmPlate Founder

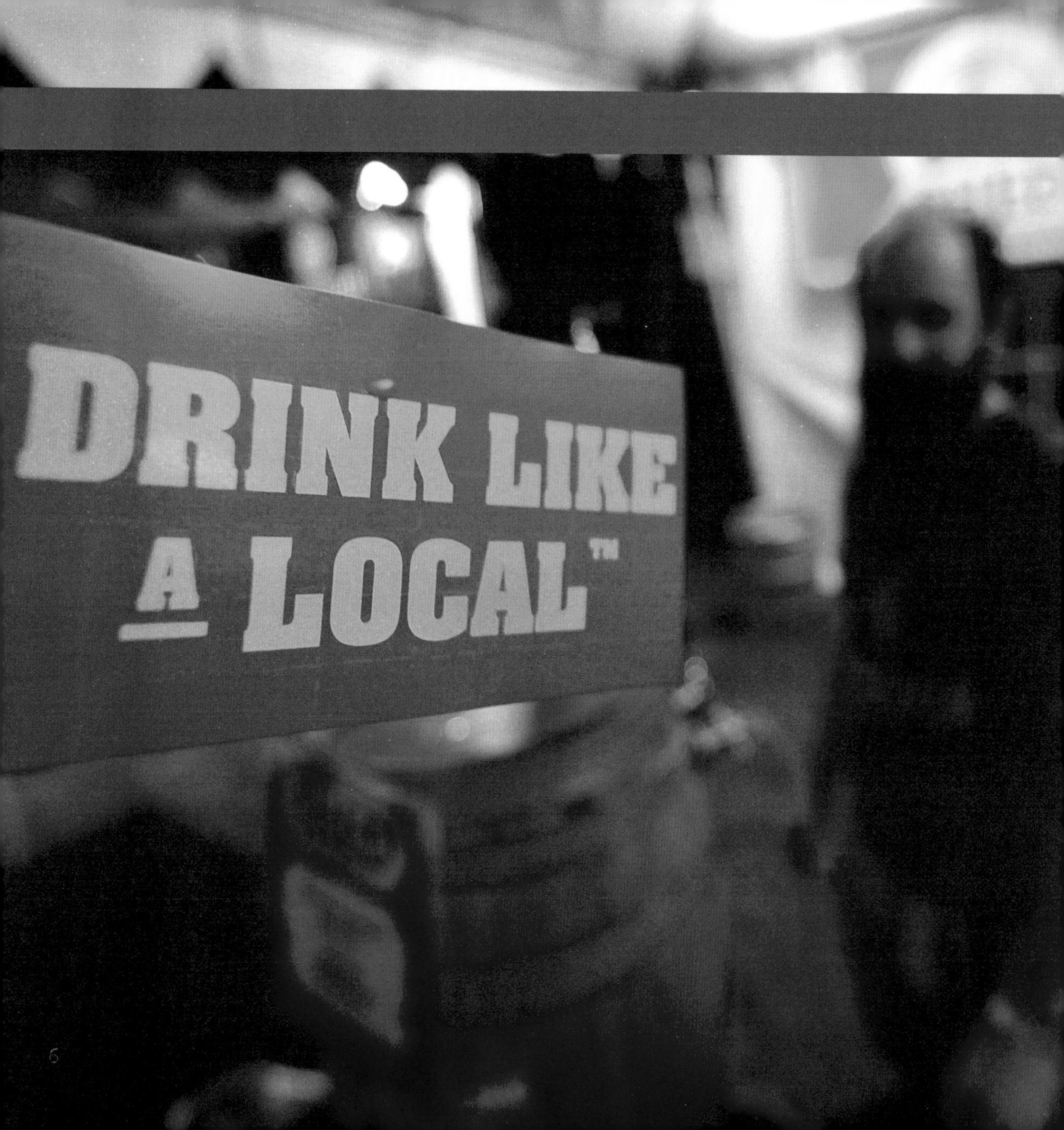
DRINK LIKE
A LOCAL™

CONTENTS

INTRODUCTION

I remember my first sip—it was a Catamount Amber Ale from Vermont's first craft brewery, and one of the first microbreweries anywhere in the United States. It was the mid-1980s and craft beer was a fringe thing—it was a missionary-like movement brought to be by a few renegades who had traveled abroad and discovered that beer could be different, very different, from what we'd all grown up with. It was a revelation.

Vermont has its share of craft beer pioneers, notably the late Greg Noonan, who followed Catamount's example. These early breweries were often cobbled together with used dairy equipment. No surprise in a state famous for its creameries, not to mention ice cream. Also not surprising was the desire to repurpose what was already at hand in typical hardscrabble Yankee fashion. Waste not.

From there things moved forward in classic northern New England style—measured, purposeful, but in no particular hurry. A return to the land of sorts, applied to brewing. These were not breweries born of grand business plans but driven by passion and a sense of place. Some grew in spite of themselves, with brewing good beer their only real goal. A few expanded, perhaps grudgingly, while others stayed small by design. Fame and fortune were not motivators; good beer was.

As time went by and craft beer grew to become a household word, Vermont soldiered on and quietly added to its diverse collection of craft breweries. Today, Vermont is home to some of the most sought-after craft beers in the country, many of which seem to unknowingly break new ground. So good . . . if you can get them! A few are deliciously rare and hard to find (literally take the dirt road to the right and then turn left at the tractor), only adding to their mystique.

> "Today, Vermont is home to some of the most sought-after craft beers . . ."

Craft beer continues to grow and has now made inroads into the American mainstream—in part due to the localvore movement as well as a growing appreciation of what's old is new again. To which Vermonters can only respond, "New? Hasn't it always been this way?"

PHIL MARKOWSKI
Brewmaster, Two Roads Brewing Co.

CRAFT BREWERS

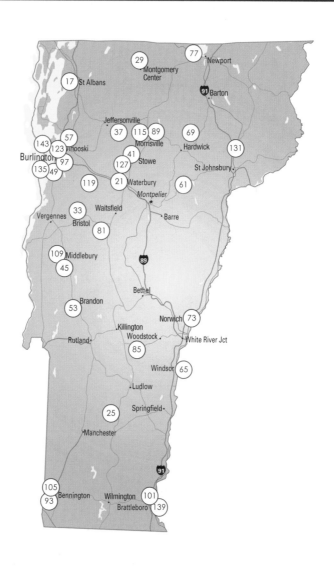

CRAFT BREWERS

KEY   Tastings   Tours/Viewing Area   Food   Beer to Go   Gear

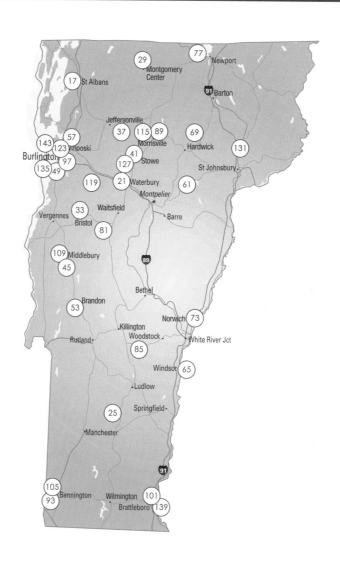

CRAFT BREWERS

KEY Tastings Tours/Viewing Area Food Beer to Go Gear

14ᵀᴴ STAR BREWING CO.

Vermont's veteran-owned craft brewery

14th Star Brewing started as an idea, and an ideal, in the mountains of eastern Afghanistan in 2010. Founders Steve Gagner and Matt Kehaya spent their precious little downtime dreaming of a post-Army life. Ten months after returning home, the two homebrewers and friends put their plan into action and leased an auto repair garage in St. Albans. Starting with one-barrel batches, their beer was soon selling out and forcing them to steadily increase production to its current level within just a year. Dedicated to quality and service, 14th Star is proud to be a part of Vermont's vibrant craft brewing scene.

Annual Production: 800 barrels

Fun Fact: Demand for 14th Star Brewing's beer is so great, the brewery plans to expand into a 15,500-square-foot space in the former downtown St. Albans bowling alley.

Don't Forget: 14th Star gives discounts at their tasting room to veterans (past and present), police, firefighters and EMTs as a small thank you for their commitment and service.

41 Lower Newton Street
St. Albans, Vermont

- - - - -

802.393.1459

- - - - -

www.14thstarbrewing.com

LEARN MORE ON TAP&MAP.COM

STEVE GAGNER | *14th Star Brewing Co.*

MASH TUN

What led you into the world of brewing?

I started homebrewing after returning from Iraq and the Army stationed me hours away from home. Instead of spending several hours a day on the road, it was more economical to rent an apartment and come home only on weekends. This was difficult with a wife and newborn at home, and I turned to hobbies to take my mind off the loneliness. I started homebrewing, and from the first beer I made, I was hooked. Every facet of the process fascinated me. This led me to share my newfound passion with my friend Matt, and we began brewing together regularly. When we started talking about retiring from the Army, we decided running a brewery would be challenging and fun. So far, so good.

What's inspired your recipes and what are you experimenting with today?

When I started, I was simply trying to imitate other phenomenal beers I enjoyed. Over the last several years, I've been really inspired by the various tastes and smells of new varieties of hops and how to manipulate them to give me what I'm looking for. I like to craft clean, drinkable, broadly appealing beers that showcase these complexities—from earthy, piney and dank to light and citrusy.

What do you think the future holds for you as a brewer and your beer as a brand?

We're expanding only as demand requires and increasing our presence throughout northern Vermont slowly. We are a brand that is here to stay and aim to fully serve the northwestern quarter of the state in the coming years.

Your ideal day (involving beer of course!) . . .

I would say my ideal day would start with breakfast with my family before sending the children off to school, brewing something new that I've had in my head and making it home in time to watch my children's sports practices and games before dinner.

"I'm on a mission to dial in every beer we brew..."

What intrigues, inspires or mystifies you?

The millions of variables in brewing that separate the phenomenal from the great brewers—grist crush, mash temps, hop schedule, yeast strain, fermenting temp, time, handling methods—all of these items have up to several hundred nuances that can completely change the beer. I'm on a mission to dial in every beer we brew to optimize what it can be.

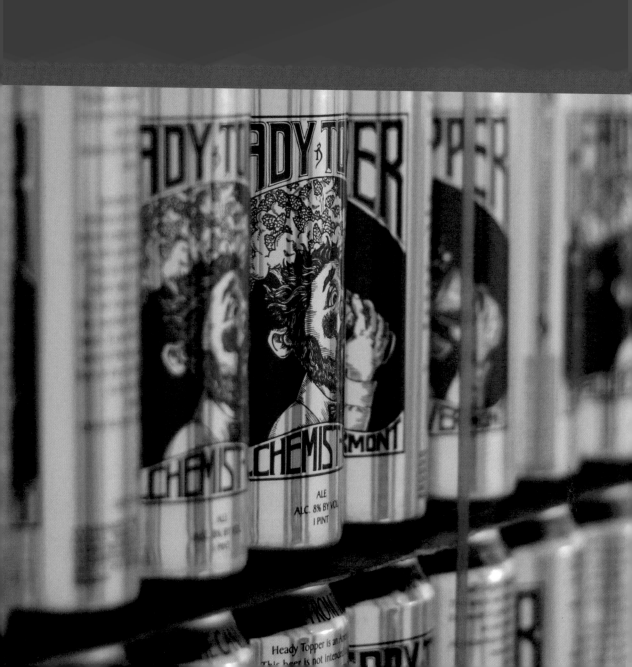

THE ALCHEMIST

Specializing in fresh, unfiltered IPA

For eight years, Heady Topper was served on tap exclusively at The Alchemist, John and Jen Kimmich's seven-barrel brewpub in downtown Waterbury. In early 2011, they built the Alchemist Cannery—a 15-barrel brewery and canning line—for the production of their flagship IPA. The first cans of Heady Topper rolled off the line just two days after Tropical Storm Irene demolished The Alchemist Pub & Brewery. Today, they brew 180 barrels a week in twelve 15-barrel batches. They never hold back inventory and move all their beer weekly in a concerted effort to provide the freshest, hoppiest packaged IPA on the market.

Annual Production: 9,000 barrels

Fun Fact: Regulars at the bars in downtown Waterbury will occasionally find one-off batches of specialty beers available on draft. These special releases go unannounced and are usually gone within a few days. Recipe choices are drawn from an extensive archive of beers developed at the original Alchemist Pub. They tend to lean heavily toward IPAs.

Don't Forget: Heady Topper usually sells out within hours of hitting the shelves. For an updated list of retailers and delivery days, refer to the Where to Buy page on The Alchemist's website.

35 Crossroad
Waterbury, Vermont

- - - - -

802.244.7744

- - - - -

www.alchemistbeer.com

Q&A

What led you into the world of brewing?

My brother-in-law. I was still in college, and I was home for a weekend and found a copy of one of his old homebrewing books. It just caught my eye. Two weeks later, he bought ingredients, we brewed a batch of beer and I just fell in love with it. It just kinda clicked because I was a business major and I didn't know what the hell I wanted to do. I just wanted to have my own business and stumbled into brewing.

What's inspired your recipes and what are you experimenting with today?

Well, recipe for now. At The Alchemist Pub, there were 70 different beers that I brewed over the years. But what inspired my recipe for Heady Topper? Same thing that inspires all my IPAs—just hops, hops, hops. I love them and I always have. As far as what we're experimenting with today, we do, even though we're only making one beer here, play around quite a bit—just different techniques, procedures, quantities of the hops. I am never 100 percent set on the recipe—I'm always open to trying something else in there.

What do you think the future holds for you as a brewer and your beer as a brand?

Hopefully more of what we're experiencing right now. We consider ourselves to be unbelievably fortunate. It is not to say that luck got us here—I think any successful business is a mix of a ton of hard work and a dose of luck. The future is just better and better. Heady Topper will only get better. When I was out West, the whole point was to meet growers. So now I will have total control—I will know the guy growing my hops, I will know the acreage it's on. In terms of the brewery, we're huge environmentalists, and the goal is to eventually be a zero-waste brewery. We have plans to recapture all the CO_2 we produce during fermentation, and we could eventually become self-sufficient and be producing all the CO_2 we use.

> # Hops, hops, hops. I love them and I always have. "

Your ideal day (involving beer of course!) . . .

I'm living it. You could say drinking beer on a mountainside or something like that. But really it's to be making a living supporting our family, the families of our employees and helping to support the families of our accounts.

What intrigues, inspires or mystifies you?

My wife. All three of those things! If it wasn't for her, we'd be out of business—really. I could have made the best beer in the world, but who the hell would have sold it? I do not have the capacity to run a business like this without her.

BACKACRE BEERMAKERS

Blenders focused on a single sour golden ale

Not a brewery in the common sense of the word, Backacre Beermakers may be the first example of a blendery in the United States. They work with nearby brewers to produce wort (unfermented beer) that is then barreled for its long transformation into beer. From this stock they carefully blend a single product: Backacre Sour Golden Ale.

Annual Production: 20 barrels

Fun Fact: The idea of making wort at a central brewhouse and fermenting it in other places is a very old one, and it is still practiced in a few areas of Belgium and Germany.

Don't Forget: Due to the very small size of the blendery, no visits or tours are offered.

Weston, Vermont

- - - - -

info@backacrebeermakers.com

- - - - -

www.backacrebeermakers.com

LEARN MORE AT PAMPLATE.COM

Q&A

What led you into the world of brewing?

Like so many American producers, we started by brewing beer at home. At some point we started to focus on making beer in a very old and traditional way, fermenting in oak barrels for months or even years with a mix of different organisms. This is different from most modern beers, even craft beers, made with a single yeast strain and fermented in sterilized steel for days or weeks. We found some oak barrels, experimented with different strains and mixtures of yeast, and got really deep into the science. After we figured out how to make beer we really enjoyed, commercial production just started to make sense.

What's inspired your recipes and what are you experimenting with today?

We make only one beer, a sour golden ale. This recipe and process are inspired by some Belgian beers that are fermented with a mix of organisms. This "mixed-fermentation" gives some extra dimensions of flavor, similar to the way that additional organisms give special character to sourdough bread or blue cheese. All our efforts, and any experiments we do, are focused on making this one beer as well as we can possibly make it.

What do you think the future holds for you as a brewer and your beer as a brand?

Backacre is extremely small, with very local distribution. We don't see this changing any time soon.

Your ideal day (involving beer of course!) . . .

Get into the blendery at nine in the morning, with the sun shining outside. We pump five barrels of beer into the blending tank, ready for bottling the next day. The blend tastes exactly as it did when we selected it days before. Amazingly, the contraption we built to clean barrels actually works this time without covering anyone in yeast. We refill the casks with sweet-smelling wort that we picked up yesterday. An army of river otters shows up to clean the equipment we've dirtied. It's still early afternoon, so there's time for a beer or two on the porch before dinner.

> "[We] focus on making beer in a very old and traditional way."

What intrigues, inspires or mystifies you?

Despite serious academic research and centuries of practical experience, so many questions about this kind of beer remain unanswered. The process is so complex that it might resist industrialization for many more decades, maybe even centuries.

BLACK LANTERN BREWING COMPANY

Small-batch, handcrafted beer

The Black Lantern Brewing Company is a fledgling brewpub started in June of 2013. The operation, which is part of The Black Lantern Inn in historic Montgomery, brings the property full circle to when is was established as an inn and tavern in 1803. The proprietors at that time would have served fresh, locally brewed beer out of wooden kegs to weary travelers, salesmen and sawmill workers alike. Just as they did more than 200 years ago, Black Lantern now brews small batches of high-quality, hand-crafted beer, favoring traditional styles and recipes. They produce classic session beers that harken back to a time gone by—a simpler time.

Annual Production: Less than 50 barrels

Fun Fact: The sawmills are long gone, but Montgomery is still a quintessential Vermont town nestled at the foot of Jay Peak. Today, the town's population is about half of what it was in the late 1800s, but it has maintained a strong sense of community and a vibrant social scene.

Don't Forget: The Black Lantern Inn is the only place you'll find their beer, along with three dozen varieties of Irish whiskey!

See page 160 for info on the brewpub.

2057 North Main Street
Montgomery, Vermont

- - - - -

802.326.3269

- - - - -

www.theblacklanterninn.com

LEARN MORE ON FARMPLATE.COM

What led you into the world of brewing?

When I relocated to the area full time, I learned the art and craft of homebrewing from my cousin, who had been brewing himself for years. I loved it from the first boil. A few years later, I purchased The Black Lantern Inn and it seemed the perfect opportunity to combine a great hobby with the business. Now I'm basically a homebrewer operating a brewpub, and I couldn't be happier.

What's inspired your recipes and what are you experimenting with today?

I'm a beer drinker. I enjoy the fellowship and society of drinking beer. My philosophy is to focus on traditional beer, faithful to historic styles, and to create session beers that promote the social experience. I tend to stay away from high-ABV beers and those with exotic and extreme flavors. They're just not my style—fads come and go but classic never goes out of style. Every day is an experiment for us. We are still in the process of selecting our house brews, those that will be available on a regular basis, and so we are constantly trying new recipes, relying largely on feedback from our regular patrons for direction. Our biggest challenge thus far has been keeping up with demand!

What do you think the future holds for you as a brewer and your beer as a brand?

I hope to grow slowly and continue to meet the demand of the local beer-drinking public here in our brewpub, one pint of beer at a time. I think if we establish our brand as a reliable source of traditional beer styles, the rest will take care of itself. We're in no hurry to change.

> "My philosophy is to focus on traditional beer, faithful to historic styles."

Your ideal day (involving beer of course!) . . .

Is there a day not involving beer? Every day is ideal for me as I build this business around craft beer. Guess I'm living the dream!

What intrigues, inspires or mystifies you?

Human behavior, as contemplated over a cold beer, intrigues, inspires and mystifies me. My previous career was in finance, in which behavior is predictable. It's been both challenging and amusing trying to figure out behavior in the service industry!

BOBCAT CAFE & BREWERY

Contemporary comfort food + house-brewed beer

Located at the base of the Green Mountains, the Bobcat Café & Brewery serves handcrafted beers alongside fresh, creative homemade food. The brewery is located in the basement of the cafe, so you can sample some of the delicious locally inspired fare while enjoying one of the Bobcat's own 10-plus beers on tap. Brewmaster Mark Magiera uses high-quality grains, hops and yeast strains to create up to a dozen true-to-style beers on a rotating basis. Building off his experience and expertise, Mark is passionate and innovative without losing the tradition behind each style of beer.

Annual Production: 260 barrels

Fun Fact: When crafting the 30 different brews offered throughout the year, Mark is known to use locally grown hops, barley and wheat.

Don't Forget: The brewery and cafe open at 4 p.m. daily, seven days a week.

See page 163 for info on the café.

5 Main Street
Bristol, Vermont

- - - - -

802.453.3311

- - - - -

www.bobcatcafe.com

Q&A

What led you into the world of brewing?
While exploring the idea of establishing a hard cider company with my father and grandfather, I started working with a brewmaster to refine my beverage production skills. In short time, he gave me some advice—to forget about cider as I had a natural ability to craft beer. I took his advice and never looked back. (Though I still like a good dry cider from time to time!) Also, it seems that working with grain and yeast is in my genes. My great-grandfather used to malt his own barley in the wood-fired stove still warm from cooking the Sunday dinner after church. Being that it was The Lord's Day of Rest, he would kiln-dry the barley on baking sheets so he could brew himself some beer at a later date!

What's inspired your recipes and what are you experimenting with today?
The concept of *terroir* is what has drawn me to recreate different styles of ales and lagers from around the world. It is not only a model for viticulture in my humble opinion. The challenges as a brewing chemist are to pull it all together (regarding ingredients) and to offer a representation of a specific style for the consumer.

What do you think the future holds for you as a brewer and your beer as a brand?
I don't know the future, however, I have a high regard for hope. Hope that people will continue to want the beer I make; hope that I will have the continued ability to make beer tomorrow and beyond . . .

> " Terroir . . . is not only a model for viticulture. "

Your ideal day (involving beer of course!) . . .
My ideal day is when everything goes right!

What intrigues, inspires or mystifies you?
I know a gentleman who travels the country on a regular basis for work. One day he sat next to me over a beer and told me that wherever he travels, he makes it a point to try the local brew. What makes this moment so memorable and inspirational is his humble comment that whenever he has that local brew in the far-off place, he compares it to something I've made. The experience makes him long for home . . .

LEARN MORE AT PABPLATE.COM

BREWSTER RIVER PUB & BREWERY

Locally known for good food, entertainment + handcrafted beer

With more than 40 years of restaurant experience between them, Brewster River Pub & Brewery owners Heather and Billy Mossinghoff have created a casual brewpub that shows off their skills and experience. Billy also has 17 years of homebrewing under his belt, so it was only fitting that they add a small-batch brewery to experiment and have some fun with! Their mission: to create homemade foods using local ingredients and to brew delicious handcrafted beer.

Annual Production: Less than 100 barrels

Fun Fact: With a 20-gallon small-batch brewing capacity, we are able to experiment with many different styles.

Don't Forget: After a hard day of skiing or riding at nearby Smugglers' Notch Ski Resort, there isn't a better place to relax with some good food and great beer.

See page 164 for info on the pub.

4087 Route 108 South
Jeffersonville, Vermont

- - - - -

802.644.6366

- - - - -

www.brewsterriverpubnbrewery.com

LEARN MORE ON FARMPLATE.COM

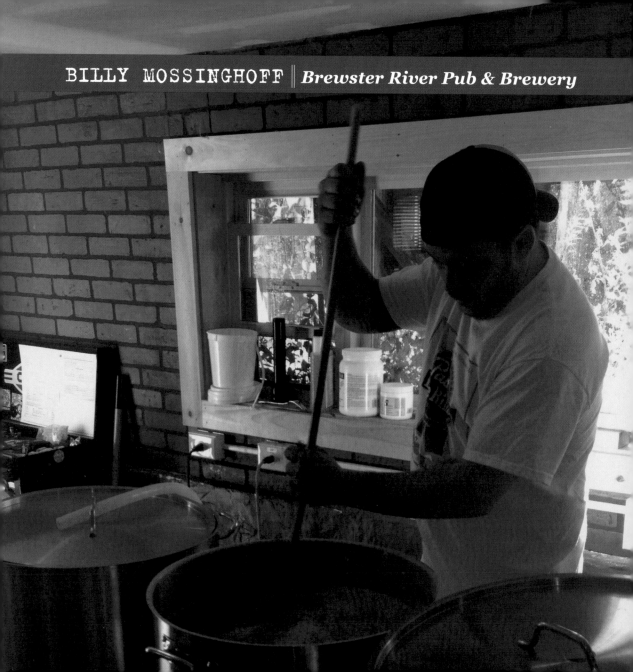

Q&A

What led you into the world of brewing?
Honestly back in my college days, I was working at a small brewpub in Pittsburgh, Pennsylvania. Picking the brain of the brewer there, I started extract brewing and found that it was a tasty alternative to drinking cheap beer. We could brew an unbelievably tasty beer for half the cost or the same cost of the cheaper brands. We won't name those though.

What's inspired your recipes and what are you experimenting with today?
The inspiration for our recipes comes from years of experimental brewing. Going from years of extract brewing to all-grain brewing allows you to make many different colors of wort, which extract brewing can hinder. That's when I really got excited and just started brewing five-gallon all-grain batches in my very small kitchen! Not good if you have a wife (girlfriend at the time)! After being kicked out of the kitchen, I built myself a 10-gallon brewery I could use in the garage. I still have that setup. It's just the constant learning and experimenting that inspires me. Being a small-batch brewery, we are always trying different styles and using different yeast strains, malts and sugars.

What do you think the future holds for you as a brewer and your beer as a brand?
I hope to mingle with more Vermont brewers and learn the art of commercial brewing. I have yet to really see that process all the way through to bottling or kegging. We definitely plan to expand and be Jeffersonville's go-to spot for craft beer, as well as distributing to the surrounding areas.

> " It's just the constant learning and experimenting that inspires me. "

Your ideal day (involving beer of course!) . . .
My ideal day would be to visit some of the great brewers of New England—spend a day with each one of them, learning the logistics of production brewing and taking in all the many different methods they use. A great ending to that day would be to sample the finished product!

What intrigues, inspires or mystifies you?
The infinite types of yeast and ingredients to choose from. The many different processes you can use to manipulate those ingredients to make an unbelievable beer. The uses of minerals to manipulate water to make different styles brewed around the world. I don't believe anyone can master brewing. One cannot stop learning.

CROP BISTRO & BREWERY

Great beers across the style spectrum

Crop Brewery's iconic location is a familiar spot to many. Since 1965, it housed The Shed—a restaurant and brewery that was part of the Stowe ski scene for more than four decades. In honor of this legacy, Crop installed a modern three-vessel brewhouse custom-designed by Caspary and started brewing in January 2013. Brewmaster Will Gilson continues to experiment with a wide variety of beers, from traditional German styles to new American classics. Eight fermenting tanks control the transformation from wort to beer, and matching serving tanks in a cold room bring the beer to tap. The brewpub is committed to the concept of matching beer styles with appropriate glassware, so see the difference for yourself and order up one of their delicious beers on tap!

Annual Production: 1,700 barrels

Fun Fact: Crop Brewery maintains over four yeast strains in-house, which makes for varied and eclectic beer offerings.

Don't Forget: Many of Crop's beers are packaged in bombers and are sold to go from their cooler. Growler fills are also available, and local enthusiasts can join their Mug Club!

See page 168 for info on the bistro.

1859 Mountain Road
Stowe, Vermont

- - - - -

802.253.4765

- - - - -

www.cropvt.com

LEARN MORE ON FARMPLATE.COM

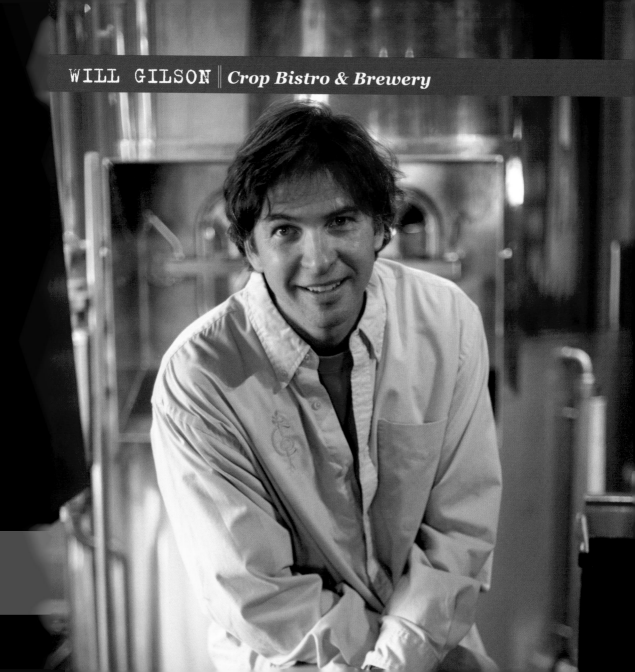

Q&A

What led you into the world of brewing?

Back in the 1980s, my buddy Stew and I spent a lot of time skiing untracked powder in the Wasatch Mountains of Utah. When the conditions were marginal, we would dedicate the day to brewing. We were all grain brewing in the late '80s. While looking for a better grain deal than one pound at a time, I started talking to a local brewery and they asked me to interview for a working brewer spot. That got my foot in the door and many more doors have opened since that fateful (and lucky) day!

What's inspired your recipes and what are you experimenting with today?

People often say the world is your oyster. I am not planning to brew with them, oysters or seaweed that is, however part of the joy of a brewpub setting is the day-to-day experimentation that tends to be limited in a production brewery. That is why collaboration beers are so fun—two great brewing minds make for some unique creations. The position I walked into—running a state-of-the-art Bavarian brewhouse—leaves me with only one limitation, my imagination. My theory has been to give people what they want. In a brewpub setting, people have a chance to actually meet the brewer one-on-one and be a part of the process.

What do you think the future holds for you as a brewer and your beer as a brand?

As for the future, we plan on dialing in our beers, which means embracing the popular ones and offering those for off-premise sale. The art and science of the brewing process comes to play as you fine-tune what your brewery produces. To grow, you need to constantly streamline the process and reinvest in your operation. Personally, I have a blast becoming familiar with everything our state-of-the-art brewery can do and sharing that excitement with onlookers in the pub.

Your ideal day (involving beer of course!)

My ideal day would involve being outdoors experiencing the awe of nature, then clinking a toast with family and friends with one of my beers.

What intrigues, inspires or mystifies you?

Foraging and finding a new patch of wild mushrooms brings out the little kid in me—mystified and full of wonder.

> "The position I walked into... leaves me with only one limitation, my imagination."

LEARN MORE OF FARMPLATE.COM

DROP-IN BREWING COMPANY

Worldly beers with Vermont character

Opening their own brewery has been a dream for Steve Parkes and Christine McKeever since purchasing in 1999 the prestigious American Brewers Guild, which specializes in intensive brewing science and engineering courses for professional brewers. Their dream became a reality when they acquired an old plumbing supply building in Middlebury and installed a brand-new Newlands 15-barrel brewing system. With more than 30 years of commercial brewing experience under his belt, Steve's passion and experience are evident in each of his beers. As a kegging-only brewery, customers can try free tastings, take home any of three different-sized growlers or enjoy Drop-In beers on tap at establishments throughout Vermont.

Annual Production: 750 barrels

Fun Fact: The ambiance within the tasting room is that of a small museum, with a collection of brewery memorabilia from throughout the United States on display. Be sure to leave enough time to browse the collection while enjoying the tastings.

Don't Forget: While no food is served at the Drop-In taproom, buy some beer, walk three paces next door and order up lunch or dinner (on select nights) at The Grapevine Grille.

610 Route 7 South
Middlebury, Vermont

- - - - -

802.989.7414

- - - - -

www.dropinbrewing.com

Q&A

What led you into the world of brewing?
I was in high school in Scotland looking at careers and I came across Heriot-Watt University, which offered a degree in brewing. I met with the professor to learn more and never looked back.

What's inspired your recipes and what are you experimenting with today?
I learned my trade making cask beers in England and so developed an understanding of the concept of drinkability. It's something I've always kept in the forefront of my mind while brewing beers here in America. While the public is evolving in terms of what it will enjoy, particularly intensely flavored beers, the premise that a beer should be clean and balanced still holds true.

What do you think the future holds for you as a brewer and your beer as a brand?
Drop-In Brewing uses the production facility owned by The American Brewers Guild Brewing School. As such, it is there to serve the needs of the students and indulge my own brewing dreams. We'll stay where we are, and the size we are, and will continue to educate professional brewers and the general public about what delicious beer is all about.

Your ideal day (involving beer of course!) . . .
I really enjoy sitting on a beach with a cooler of National Bohemia cans beside me. However, beer festivals can be a lot of fun too. Judging at the World Beer Cup with judges from all over the world is an amazing experience. But graduation day at the end of an American Brewers Guild class probably tops them all. You cannot be around that level of passion, optimism and sense of accomplishment and not be inspired by it.

> " I'm inspired by the new generation of brewers. "

What intrigues, inspires or mystifies you?
I'm intrigued by what will happen next. I'm inspired by the new generation of brewers who are going to make it happen. I'm mystified by what is going on in the minds of people who review beers on the internet.

FOR SALE

FIDDLEHEAD
BREWING COMPANY

SHELBURNE, VERMONT

DDLEHEAD
BRWING COMPANY

FIDDLEHEAD BREWING COMPANY

Be a fern believer

Fiddlehead Brewing Company is located in the beautiful town of Shelburne, Vermont, just south of bustling Burlington. Their mission is to produce full-flavored beers with the true beer connoisseur in mind. They focus on depth of flavor and work to incorporate local ingredients whenever possible. Renowned brewmaster and owner Matthew Cohen (known industry-wide as Matty O) is on a continual quest to craft the perfect pint. While the brewery always has several seasonal selections available on tap for free samples and growler fills, their flagship beer—the Fiddlehead IPA—can be found on draft lines all over Vermont.

Annual Production: 4,000 barrels

Fun Fact: The brewery taps a new beer every two weeks.

Don't Forget: To support your local brewery, and neighbhood pizzeria! Grab a growler then head next door for delicious wood-fired pizza at Folino's *(see page 172)*.

6305 Shelburne Road
Shelburne, Vermont

- - - - -

802.399.2994

- - - - -

www.fiddleheadbrewing.com

LEARN MORE ON PAMPLATE.COM

What led you into the world of brewing?

When I entered college, I was unsure of my calling. I took a number of different classes and found myself most interested in anthropology. I loved learning about different cultures and peoples. One thing I quickly realized was that almost every culture had some sort of fermented beverage. I was fascinated by these different drinks and set out trying to produce some examples. What began as a study of culture turned into a lifelong exploration into the amazing world of fermentation science.

What's inspired your recipes and what are you experimenting with today?

When developing a new beer, I look for inspiration from a number of different sources. With the Internet, there is so much information readily available to brewers that this is usually the first place I look. People have been brewing for thousands of years and experimenting with all types of different raw materials. Chances are if you think up an idea, it's not original. Someone somewhere has tried to brew it before, and through that person's trial and error, a brewer can learn a lot. Many of my experimentations begin with extensive research combined with many years of professional experience. I'm always looking to push the boundaries by using a variety of different raw materials in an attempt to create a pleasurable and unique drinking experience.

What do you think the future holds for you as a brewer and your beer as a brand?

It is currently a very exciting time for craft beer. Breweries are opening at a blinding pace, and as a brand, I see the challenge in the future being trying to stand out in a crowded field. Fiddlehead's goal has been the same since we opened, and that goal is to remain loyal to our local customers and provide them with an amazing drinking experience at a fair and competitive price.

Your ideal day (involving beer of course!) . . .

When I'm not in the brewhouse, my ideal day consists of family, good friends and great beer. For every experience in life there exists the perfect beer to pair it with. Set out to discover yours.

> ## "I'm always looking to push the boundaries."

What intrigues, inspires or mystifies you?

Brewing has been somewhat glamorized in our culture. The reality can be much different—long, hard, cold, wet days with your life tied so closely to the beer. To me what makes it all worth it is seeing the smiles on customers' faces when they really enjoy our beer, or hearing from a couple, as I once did, that they must have Fiddlehead at their wedding! To be partly responsible for bringing people joy—that's what makes it all worth the effort.

FOLEY BROTHERS BREWING

Do it the hard way

Foley Brothers Brewing produces handcrafted beers highlighting Vermont ingredients that are distinctive and easy to drink. Brothers Patrick and Daniel Foley realized their dream of starting a brewery in November 2012. They are dedicated to selecting the finest ingredients and operate with diligent brewing techniques, resulting in beers that taste just how they think beer should taste. Their motto "Do it the hard way" exemplifies their belief that there are no shortcuts to creating artisanal beers that are deserving of the family name. Their first "Native" line of beers includes Native Ginger Wheat Ale, Native Brown Ale and Native IPA.

Annual Production: 700 barrels

Fun Fact: The original Gaelic form of Foley is *O Foghladha*, derived from the word *Foghladha*, which means "pirate" or "plunderer"—hence the gnarly pirate graphic on some of their special brews, courtesy of their friend and artist Rosy Metcalfe.

Don't Forget: Foley Brothers beers are sold in 22-ounce bottles as well as on draft throughout Vermont. You can also fill up growlers at their tasting room, housed in a classic red barn on five picturesque acres with an on-site B&B run by their family. The barn is also home to Neshobe River Winery, another family business.

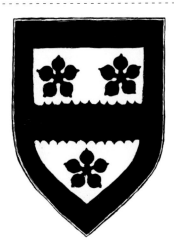

79 Stone Mill Dam Road
Brandon, Vermont

- - - - -

802.247.8002

- - - - -

www.facebook.com/FoleyBrothersBrewing

DANIEL A. FOLEY | *Foley Brothers Brewing*

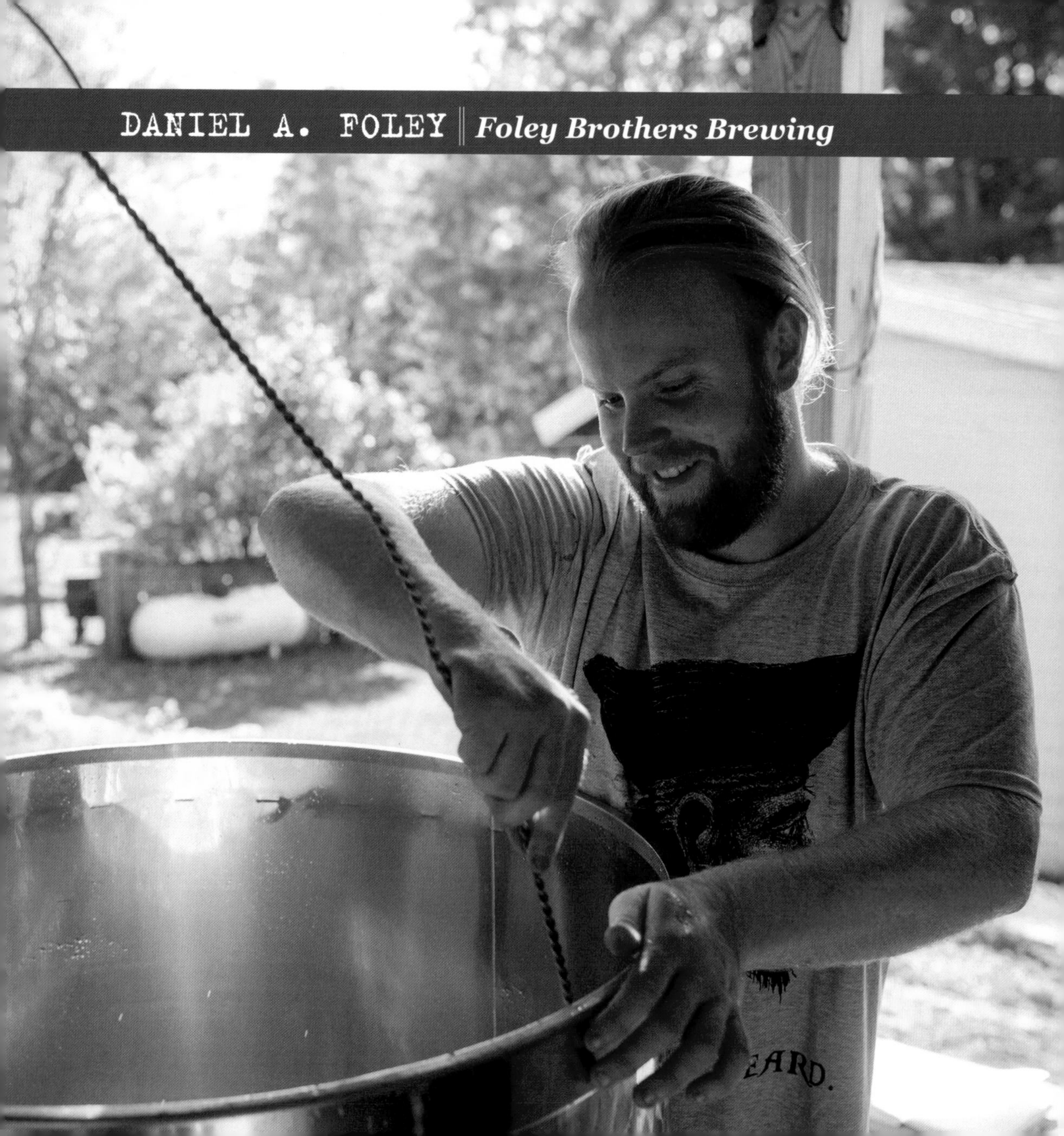

Q&A

What led you into the world of brewing?

I have a degree in biology and have always been fascinated by the natural world and the art and science of cooking. Years ago, I decided to experiment with brewing a batch of beer. I fell in love with the process, and obviously the end product isn't so bad either. It's a perfect fit for all of my interests and the best part is I get to share the results with the important people in my life. Now as a brewmaster for Foley Brothers Brewing, I get to share all of that with a larger audience, and it's really a dream come true.

What's inspired your recipes and what are you experimenting with today?

When I began putting recipes together, I started with the styles of beer that I like to drink and modified them based on how I wanted them to taste. Many of the ingredients, like maple syrup and ginger, are some of my favorite things to cook with, so I naturally used them as adjuncts. From there, it was one trial after another. I'm currently experimenting with my first saison and aging another beer in whiskey barrels.

What do you think the future holds for you as a brewer and your beer as a brand?

I'm lucky to have the opportunity to make exactly the kinds of beers my brother and I want to make. We're close to our customers and try to offer the styles they ask for, with the Foley Brothers signature. We're currently working on doubling our production and look forward to bringing on more employees to help us achieve our vision for the brewery.

> " I am inspired by the fermentation process. "

Your ideal day (involving beer of course!) . . .

Go for a run, kiss my family goodbye and get to the brewery early. Brew a batch of beer, and maybe bottle a small test batch I'm working on. Have the whole family and friends over for wood-fired pizzas, play a big game of wiffle ball with some good tunes on in the background, and of course drink some Foley Brothers brews. Polish it off with a bonfire and some s'mores. I call that a great day! If I could somehow fit fishing in there, that would be ideal.

What intrigues, inspires or mystifies you?

I am inspired by the fermentation process and love working with a product that is built upon a unicellular organism with such a rich history and amazing characteristics.

FOUR QUARTERS BREWING

Winooski's neighborhood brewery

Four Quarters Brewing is a small-scale production brewery three blocks from downtown Winooski with a tasting room for samples and growler fills. The brewery produces high-quality, artfully crafted beers influenced by Belgian and American traditions. Four Quarters refers to the four quarters of the moon, the four seasons, the four elements and the cycle of life. All of these are our interfaces with the natural world, where we have the opportunity to reflect its inspiration into our craft.

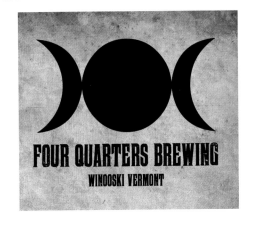

Annual Production: 400 barrels

Fun Fact: Brian is not only a brewer, but a beekeeper, gardener and an ordained minister.

Don't Forget: Limited quantities of local honey, bread, cheese and other products are sometimes available in the retail shop.

150 West Canal Street
Winooski, Vermont

- - - - -

www.fourquartersbrewing.com

BRIAN ECKERT | *Four Quarters Brewing*

Q&A

What led you into the world of brewing?
I was exposed to craft beers during a cross-country trip several years ago. Shortly after, I looked into homebrewing and was gifted a homebrew kit by my wife. My curiosity took over and brewing became an exceptional creative outlet, combining an exploration of flavors, tinkering with DIY projects and crafting anything I could dream up. That interest turned into an obsession, which led to homebrew clubs, beer judge certification, advanced brewing sciences, microbiology, welding, electrical engineering . . . the list goes on. I enjoy being able to create something tangible out of what was once in my mind. Brewing seems to have that extra dimension where it has the influence of ingredients, the equipment, the process and the artist.

What's inspired your recipes and what are you experimenting with today?
Many things provide inspiration: flavors, the seasons, people and just diving deeper into the wealth of knowledge that's out there in regards to all of these. Understanding water chemistry is an interest of mine right now, as well as wrangling spontaneous fermentation and understanding the depth of wild yeast and bacteria's influence on flavors. I love the fact that these styles of beers have living microorganisms in them and are just as beneficial as the ones you find in some foods such as yogurt, cheese, kimchi and sourdough. Not to mention they're some of the best-tasting beers in the world.

What do you think the future holds for you as a brewer and your beer as a brand?
Infinite possibilities. I'm loving all the opportunities Winooski and Vermont have to offer, and all the great ingredients that are grown here. I also love collaborating with other creative people, especially ones in other crafts. It can be a truly symbiotic process resulting in something much more special than I could ever have made by myself.

> " Brewing [is] an exceptional creative outlet... "

Your ideal day (involving beer of course!) . . .
Really any day that I can share a good beer with family and friends (it doesn't even have to be my own) and enjoy things that we can all appreciate, like good music, good food and just the general enjoyment of life.

What intrigues, inspires or mystifies you?
Trying to wrap my head around our place in the Universe. This consumes my mind most of the time. It puts everything about life in perspective, it keeps me in check, makes me appreciate the things that are most important and allows me a lot of breathing room for things that aren't so important.

GRATEFUL HANDS BREWING

Small-batch dark ales

Grateful Hands Brewing handcrafts small batches of flavorful, complex and unique ales, with an emphasis on dark ales. Owned and operated solely by a husband-wife team, it is not a large company and it has no desire to become one. Keeping the creativity and fun involved in crafting beer is the priority—after all, the hands-on creative aspect of brewing is what makes them look forward to heating up the kettles every week!

Annual Production: 65 barrels

Fun Fact: Owner-brewer Ricky McLain is a licensed professional structural engineer with a master's degree. According to the executive director of the Vermont Brewers Association, that makes him the only brewer in the state with letters after his name—that's Ricky McLain, M.S., P.E.!

Don't Forget: Be sure to check out their website for stores where their beer is sold (all within a short distance of the brewery) as well as for hours for tastings, growler fills and bottle sales at the brewery as it is only open limited hours and seasonally.

EST. 2012
Small Batch
GRATEFUL HANDS BREWING
Dark Ales
CABOT, VT

2211 Route 2
Cabot, Vermont

- - - - -

802.249.4092

- - - - -

www.gratefulhandsbrewing.com

RICKY McLAIN | *Grateful Hands Brewing*

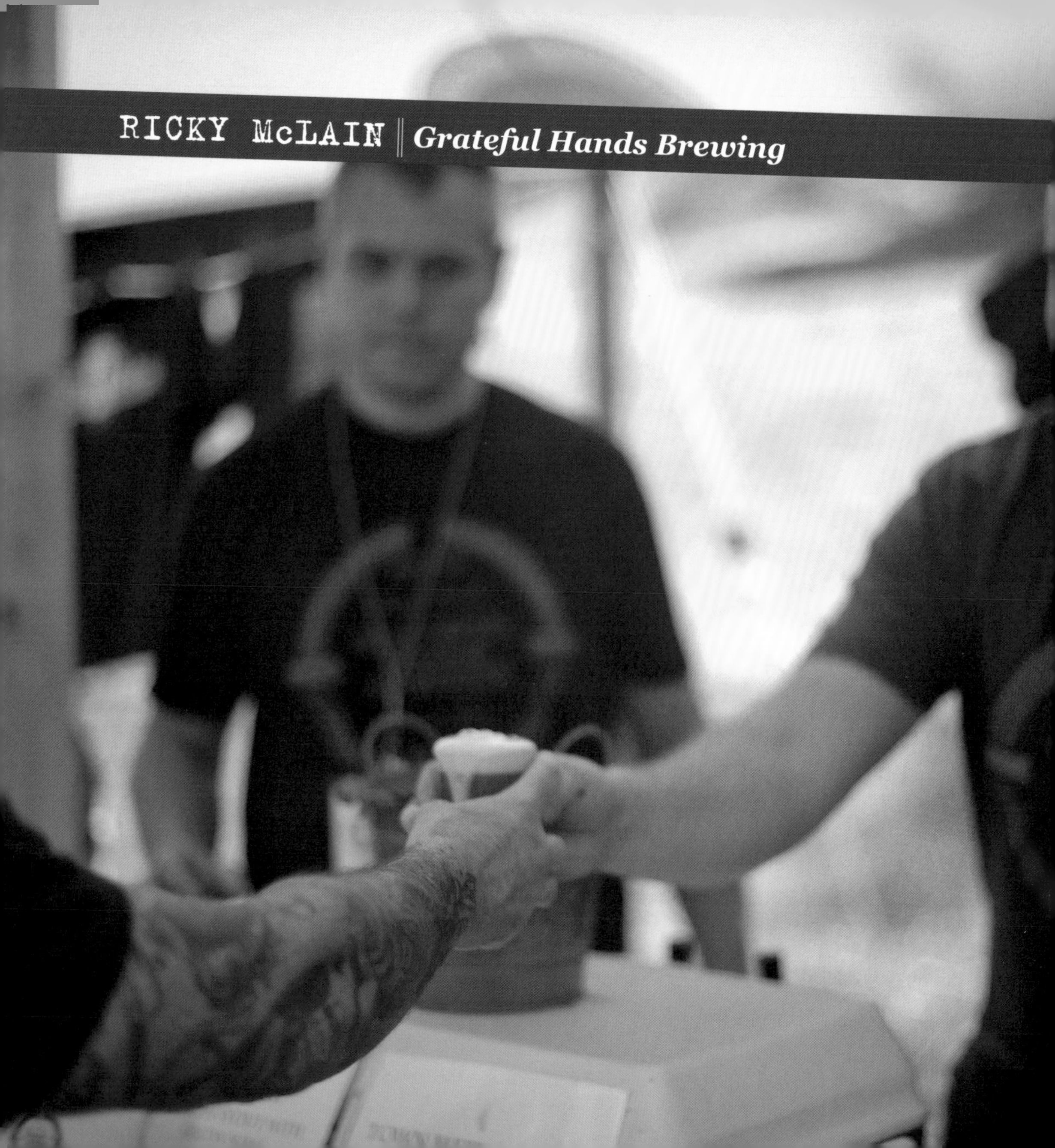

What led you into the world of brewing?

I don't know that I was led into the world of brewing; it is more that I finally discovered something so fantastic and creative that it turned into a passion that is now an integral part of my life. I'll admit that I used to be so oblivious to what craft beer was that I actually ordered a margarita on my 21st birthday! But shortly after I moved from Maine to Vermont, my wife and I bought a small homebrewing kit. That was about 3½ years ago. An intense passion for brewing and learning about processes, ingredients and all other brewing and conditioning topics ensued.

What's inspired your recipes and what are you experimenting with today?

I have gained all my brewing knowledge from books, podcasts and videos, and so I would say my recipes are inspired by ideas, flavors and concepts that I hear about through those mediums. Not copying a recipe, but hearing about an idea, modifying it in a way that I think would be better and going from there. I also like to think about beer flavors as I'm eating. Earlier this year we did a saison with fresh orange zest and fresh-squeezed orange juice using this food-beer pairing concept.

What do you think the future holds for you as a brewer and your beer as a brand?

What I really love about this boom in the number of breweries is that beer drinkers are becoming more discerning regarding flavors and quality. It pushes us to always be improving . . .

Your ideal day (involving beer of course!) . . .

It certainly includes spending time with my wife and daughter. We love to hike and explore the state through travel and outdoor adventures. We also love visiting breweries and exploring what they are focusing on. A quiet evening back at home with a Grateful Hands Common Sense Black IPA and a book in hand caps out the day.

> " I...like to think about beer flavors as I'm eating. "

What intrigues, inspires or mystifies you?

As an engineer, my mind works in a routine of identifying and finding the source of problems then coming up with solutions. Brewing certainly follows the same routine, whether it's a flavor that just seems a bit off or an issue in the brewhouse, I really enjoy the problem-solving involved. What mystifies me is the crazy amount of beer hunting/trading/scalping that goes on these days. I would love a world in which every town had its own small brewery and if you wanted to try the beer, you would plan a vacation to the area and hit up the local breweries.

HARPOON BREWERY

Love beer. Love life. Harpoon.

Harpoon has always worked hard at two things: brewing great beer and welcoming their customers to their breweries in both Vermont and Boston. They create beer styles to provide their beer-drinking friends with fresh, fun and interesting choices. Harpoon draws on numerous brewing traditions to make their beers, but they always add their own interpretation of how the styles can be best matched to suit the tastes of their own team and their loyal beer-drinking fans.

Annual Production: 200,000 barrels

Fun Fact: Harpoon's annual Kettle Cup is a brewing competition among Harpoon employees that takes place at the Windsor brewery every September. The first-place team gets to brew their winning beer for the Harpoon 100 Barrel Series.

Don't Forget: Harpoon hosts special events and live music throughout the year. Every July they host the annual Championships of New England Barbecue, and every October they roll out the barrels for their Octoberfest celebration right at the brewery. The Riverbend Taps & Beer Garden is open every day for pints of fresh Harpoon and great pub fare.

See page 174 for info on the beer garden.

336 Ruth Carney Drive
Windsor, Vermont

- - - - -

802.674.5491

- - - - -

www.harpoonbrewery.com

Q&A

What led you into the world of brewing?
I discovered what beer could be after joining the Navy in 1989 and seeing what great beers were available in Europe. Microbrewed beers soon caught my attention and I was seeking out all the great beers from the early pioneers. This, of course, led to homebrewing and ultimately to going to school where I could learn to brew beer professionally. I was hooked.

What's inspired your recipes and what are you experimenting with today?
Personal experiences determine what I like to experiment with. Sometimes it's a single ingredient like ginger or a long-lost style like Mastne. On the 2013 Company Beer Culture Trip to Poland we visited the Brackie Brewery in Cieszyn. They had revived an old recipe for the town's 1200th anniversary. It was a wonderful beer and experience. It gave me a chance to experiment with Polish hops and malt—something I may have never thought to even try.

> " I live in the mountains of Vermont and make beer for a living. The truth is, every day is my ideal day. "

What do you think the future holds for you as a brewer and your beer as a brand?
For me as a brewer, it's simple—a long and fruitful career working at Harpoon. Harpoon has been making great beers for over 25 years now, and we will continue to strive to be an iconic brewery in New England. We've had a good start.

Your ideal day (involving beer of course!) . . .
I live in the mountains of Vermont and make beer for a living. The truth is, every day is my ideal day.

What intrigues, inspires or mystifies you?
Yeast. The fact that yeast produces the only mood-altering drug with a caloric value is mystifying to me. There are a lot of other microbes we use to our benefit but none that have that same power. Some people have suggested fermentation is the reason why we took up an agricultural life in place of hunting and gathering. What I'm enjoying in my beer is also in my car's gas tank . . . that amazes me every time I think about it.

HILL FARMSTEAD BREWERY

Artistry, science, fermentation + a sense of place

Hill Farmstead Brewery is the culmination of travel and insight, of friendships and explorations and of discovering a sense of place. Upon the hand-hewn land of his forebears, Shaun Hill honors eight generations of Greensboro ancestry by thoughtfully engaging with his heritage and with the spirit of his farmstead ales. The brewery is his effort to revive, diversify and prolong the memory of the Hill Farmstead.

Annual Production: 1,800 barrels

Fun Fact: Prior to returning to his family home in Vermont, Shaun spent two years (having a lot of fun!) as head brewer at Nørrebro Bryghus, a brewpub in Copenhagen, Denmark.

Don't Forget: The Hill Farmstead retail shop is the only place their beer is regularly available by the bottle or for growler fills. Cash sales only.

403 Hill Road
Greensboro, Vermont

- - - - -

802.533.7450

- - - - -

www.hillfarmstead.com

Q&A

What led you into the world of brewing?
There was this epiphany I had one day on a balcony. I remember realizing that people were really passionate about brewing, but I am more passionate about sense of place, like the experience and the space around it, as opposed to the actual process of brewing, which I am not that enthused by. That was the moment I realized I was good at something and I could use it as a vehicle to fulfill whatever contribution to the world I wanted.

What's inspired your recipes and what are you experimenting with today?
Literally today, sour blueberry beer. We have beers aging on different fruits—that's a new direction for us. Blending is inspiring—trying to achieve new flavors using multiple threads that were fermented, or even just through happenstance, having four different beers that went through three different processes and then blending them back in different ratios . . . sort of like wine, really.

What do you think the future holds for you as a brewer and your beer as a brand?
Hopefully, reaching the end point of production (as in no more growth), then working backwards and improving efficiency and perfection. To reach a more calm moment. As far as the brand goes, if we stop making beer at 4,000 barrels, that means we are a northern Vermont operation—a small operation defined at one point in time. I don't think our beer will leave Vermont again . . . it's to the detriment of people close to home who wouldn't be able to get their beer. The brewery has a responsibility to the community that has created the brewery, so giving back, and keeping things closer to home, I think has a greater impact on the local area. It brings people here.

Your ideal day (involving beer of course!) . . .
At this point, it would be walking into a brewery which I happen to own, everything in the right place, everything perfectly clean, every beer tank shiny, and the contents of every single tank tasting perfect to me. No mold, no stagnant air, no puddles of water on the floor—just sort of an ideal, beautiful, perfect factory for beer.

> " The brewery has a responsibility to the community. "

What intrigues, inspires or mystifies you?
Fermentation still mystifies me. Music is probably most inspiring, and living here and being surrounded by beauty and serenity. That, and fermentation, and my cat Clover are all inspiring. Intrigues? I haven't had a great deal of intrigue in my life.

A MUR

Whistling Pig
Red Ale

ALE

Call us at (802) 649-1143

JASPER MURDOCK BREWERY

At the Norwich Inn

In 1993, Norwich Inn co-owner Tim Wilson started brewing small batches of beer and quickly realized his beer was a hit! To accommodate the demand, he built the current four-barrel brewhouse. Twenty years later, Jasper Murdock Brewery continues to craft quality ales and lagers from the finest malts and hops, including hops grown in their own garden! The beers today are made with the same care that first inspired Tim to start the brewery. In addition to German lagers, Belgian ales and an occasional American-influenced beer, the brewery offers many fine English and Irish ales, which the inn has developed a fine reputation for over the years. Because filtration can strip flavor and body from a beer, the brewery allows the yeast to settle naturally in an extended cold-aging period to ensure that all the goodness reaches your palate.

Annual Production: 235 barrels

Fun Fact: Beer is pumped 140 feet from the brewery's cellar, into the basement of the inn and then to the taps in the alehouse!

Don't Forget: Jasper Murdock beer can only be obtained at the alehouse and in 22-ounce bottles sold at the front desk of the inn.

See page 178 for info on the alehouse.

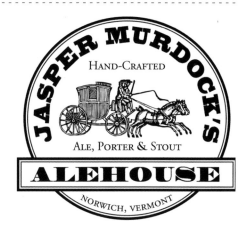

325 Main Street
Norwich, Vermont
- - - - -
802.649.1143
- - - - -
www.norwichinn.com

JEREMY HEBERT | *Jasper Murdock Brewery*

Q&A

What led you into the world of brewing?
An anthropology professor I worked for while attending UVM, Professor William Mitchell, introduced me to quality beer. He took me out to dinner at the end of a term as a thank you for my assistance. Once seated, he asked me if I'd like a beer, but when I politely declined, he looked at me with a twinkle in his eye and asked, "But have you ever had a good beer?" A couple of Munich Dunkels later, I realized that my mission that summer was to seek out "good" beers. Once I discovered beer could actually be outright delicious, I decided to research the possibility of making my own at home. My research was successful, and as they say, the rest is history.

What's inspired your recipes and what are you experimenting with today?
I love brewing to style, and primarily, that's how I brew. But I do like to experiment on occasion. In my 20-plus years of brewing, I have brewed primarily with European hops. Lately I have been working with a number of American varieties and am having a blast. I think as I have been getting older, my palate has become more accepting of bolder flavors. I guess this current chapter of my brewing career was timed perfectly!

What do you think the future holds for you as a brewer and your beer as a brand?
Since I make beer for a small brewpub with six faucets in the pub and only one year-round brew, I will continue to develop new beers, which is the part of the process I enjoy the most. Additionally, I hope we will one day offer cask ales in the pub, along with growlers and maybe even some barrel-conditioned beers. Currently I am working on tweaking the draft system to support an occasional nitro brew.

> "I love brewing to style...but I do like to experiment on occasion."

Your ideal day (involving beer of course!) . . .
My ideal day would start with a nice country breakfast. I'd then spend a few hours working in the garden with my wife and our cat Rocket. We'd have a light lunch followed by a short nap, and then go for a run. At the completion of our run, we'd break out the beer!

What intrigues, inspires or mystifies you?
I am intrigued by our cat's athletic ability. I am inspired by my wife's commitment to human services. And I am mystified by people who say that they "don't like dark beers" when after further questioning it is revealed that they have only tried one or two in their lives.

KINGDOM BREWING

That far North

KINGDOM BREWING
Newport Center   Vermont

Brian and Jennifer Cook, otherwise known as Team Cook, are brewers, farmers, horticulturists and bartenders all in one day. Their brewery, which has the distinction of being Vermont's northernmost brewery (situated just minutes from the Canadian border), is located on the family's working farm. Their focus on being "Vermont Green" is evident in many of their practices. They grow hops, fruits and vegetables as well as maintain a herd of Black Angus. The farm operates as much as possible as a sustainable ecosystem. The cows are fed spent grains, yeast and hops from the brewing process, and the brewery operates with a geothermal cooling system that reduces its energy costs to about that of a typical small residence.

Annual Production: 400 barrels

Fun Fact: Since Kingdom Brewing is almost as much a farm as a brewery, a visit is always family-friendly and there are cow feedings most weekends.

Don't Forget: The tasting room is open Thursday through Saturday from 3:30 to 6:30 p.m., and tours are conducted by appointment. The brewery has an open house every Columbus Day weekend, so make a visit to this beautiful region of Vermont part of your foliage tour.

1876 Route 105
Newport, Vermont

- - - - -

802.334.7096

- - - - -

www.kingdombrewingvt.com

LEARN MORE ON FARMPLATE.COM

Q&A

What led you into the world of brewing?
Science class in high school is where my fascination with both beer and wine started. I was fascinated by the whole process of making alcohol, so I did, and it was awful! I tried again, and it was better. I went to college, earned a degree, then helped manage a few profitable business. Several years ago, we were moving some boxes around and my wife found my brewing equipment. She asked me to show her how to make beer, and a year later I came home from work and she was on the front porch brewing with mulling spice. I said, "I don't have any beers with mulling spice in them." She answered back, "I do!" I guess that's when we decided to get more serious about brewing. We then started brewing a batch of beer a week for five years to test our recipes on friends.

What's inspired your recipes and what are you experimenting with today?
The inspiration for our recipes comes from hours of research. We look to history for our inspiration. We want to recreate styles people used to enjoy. I am into the War of 1812 200th anniversary types of beer and log camp beers. They may be funky to some, but they are loved by many.

What do you think the future holds for you as a brewer and your beer as a brand?
We are a very green brewery. We are already using the earth to cool our fermenters and wood to heat our hot water. Integrating solar hot water and energy will be the next step. We will simplify and perfect what's good and remove what's not.

Your ideal day (involving beer of course!) . . .
I think the best part of being a brewer is visiting other breweries and tasting their funky creations. I'm now trying to make beers that we and our friends and customers want to try. I don't want to be an IPA or stout drinker only. I love those beers but in the fall when our black spruce IPA hits the taproom, that's my favorite beer while it lasts. A true microbrew drinker is not loyal to one style. They try them all. I say let's try them all, but not in one day!

> "We look to history for our inspiration."

What intrigues, inspires or mystifies you?
Why people think it's such a long way to go to Newport, Vermont. In the time you spend reading this book, you could have driven to my brewery and had some great beer. Cheers!

LAWSON'S FINEST LIQUIDS

Straight from the Green Mountains to your head

Lawson's Finest is a small-batch artisanal micro-brewery located in Warren, Vermont. Their goal is to provide local brews of the highest quality and freshness while crafting unique new recipes and emulating the best of widely appreciated styles. Their beer is custom-crafted in a tiny facility tucked in the woods on Lincoln Mountain.

Annual Production: Less than 500 barrels

Fun Fact: Brewer Sean can also be found during the winter months leading snowshoe tours on the slopes of Mad River Glen, where he directs the Naturalist Program and is a telemark skier. He holds degrees in environmental studies and forestry. Co-owner and wife Karen helps with bottling and running the business.

Don't Forget: Lawson's Finest is not open to the public (no tours, tastings or retail at the brewery). Their beer is available at only a few retailers, restaurants and bars, and occasionally at the Waitsfield (summer/fall) and Montpelier (winter) farmers' markets. Bottles sell out quickly so check the delivery schedule, and plan to visit one of their draft locations as well. Lawson's Finest brews a wide variety of beers, and the selection often varies. Visit their website and blog for the latest updates.

Lawson's Finest Liquids

Straight from the Green Mountains to your Head!

Warren, Vermont

Warren, Vermont

802.272.8436

www.lawsonsfinest.com

What led you into the world of brewing?

When I was a college student at UVM, a good friend shared his homebrew with me. "Wow!" I said, "You made this?!" I asked if he would show me how to brew, and a couple weeks later we were cooking up a batch of maple wheat ale in my apartment. I was forever hooked! Several jobs at brewpubs out West further enticed me, but it took nearly 20 years of homebrewing and another career to bring me back to the idea that this could be a successful and worthy business venture.

What's inspired your recipes and what are you experimenting with today?

I love hops tremendously! Hop-forward beers and IPAs form the core of our offerings. Variety is also very important, and I strive to brew at least two dozen beers each year in a range of styles. Lawson's has helped put Vermont on the map for maple beer. Each year we brew a batch of very special Maple Tripple Ale using 100 percent concentrated sap as the base of the beer instead of water.

What do you think the future holds for you as a brewer and your beer as a brand?

A ton of hard work! Lawson's Finest has arrived where it is today through a dedication to excellence and a strong work ethic. Our vision is to create fantastic tasting beers as a local brewery in the Mad River Valley, get it to our customers as fresh as possible and create the best quality of life we can imagine. As long as Lawson's Finest continues to craft top-quality beer, people will be enthusiastic about seeking it out.

Your ideal day (involving beer of course!) . . .

My ideal day includes getting a good dose of time outdoors, pairing amazing beer with creative foods and making new connections with people over a beer. Every day is an adventure, with both challenges and rewards. My wife and business partner Karen is a key part of an ideal day, helping to guide our efforts and leading the way with parenting our two daughters.

"Brewing is an act of creation."

What intrigues, inspires or mystifies you?

Brewing is an act of creation. It took many years for me to fully appreciate the art and craft of making beer, and it finally led me to become a professional brewer. I am constantly baffled by the extraordinary effort people put into obtaining a sought-after beer. Sometimes it's exciting—other times, simply crazy! I feel very fortunate that so many people have discovered full-flavored beer and are so passionate about craft beer.

LEARN MORE AT FARMPLATE.COM

83

LONG TRAIL BREWING COMPANY

Hit the trail

Long Trail Brewing Company began its mission to brew high-quality craft beer in 1989. The first official batch was brewed in the basement of the old Bridgewater Woolen Mill. Long Trail soon outgrew the space and, in 1995, relocated just up the road to a new state-of-the-art facility designed to minimize the impact on the environment. Steam from the brewing process is condensed into hot water then used to heat water for the next brewing cycle, saving up to 2,000 gallons of propane a month! Long Trail also uses just one-third of the water of a conventional brewery. Long Trail combines the very best ingredients and environmentally minded brewing processes with a genuine passion for brewing high-quality craft beer.

Annual Production: 95,000 barrels

Fun Fact: Vermont's Green Mountain Water Environment Association named the water produced by the brewery's artesian wells "Vermont's best drinking water," lending credence to the saying, "It's in the water!"

Don't Forget: Thanks to their farmhouse-turned-pilot-brewery, Long Trail is able to offer creative new brews on a daily basis at the Visitor Center.

See page 180 for info on the brewpub.

5520 Route 4
Bridgewater Corners, Vermont

- - - - -

802.672.5011

- - - - -

www.longtrail.com

LEARN MORE ON FARMPLATE.COM

Q&A

What led you into the world of brewing?
I learned to homebrew from one of my professors in college. By the time I was a senior, I knew I wanted to make beer for a living. So I talked my way into the brewing business and have never looked back.

What's inspired your recipes and what are you experimenting with today?
Generally new beers evolve out of some new ingredient or process we want to try. We created our Maple Maibock around the maple syrup a co-worker harvests every year, and we recently brewed a white IPA featuring grapefruit rind as a twist on the more typical use of orange peel. Currently we're working with a lot of new hops and some different yeast strains, in addition to just trying new combinations of classic ingredients.

What do you think the future holds for you as a brewer and your beer as a brand?
I still feel like I have a lot to learn, even after 20-plus years as a professional brewer. We're pushing frontiers as far as the types of beers we make and how we make them, looking both forward (with new technologies) and backwards (using traditional methods like barrel aging) as we seek to grow as a company. We're also looking to reach more people. Long Trail has long been popular in New England, but we're branching out into some new markets as well.

Your ideal day (involving beer of course!) . . .
I like to start with a morning hike or hit the mountain bike and get into the brewery by 8 a.m. Ideally everything is going to plan and we can work on new projects and recipes and evaluating what's in the tanks and what's cooking in the pilot brewery. We like to taste around 11 a.m. Tasting is always the best part of the job. I try to finish everything in time for a beer in the pub at the end of the day.

> "I'm really in awe of today's craft fans."

What intrigues, inspires or mystifies you?
I'm intrigued and inspired by beer consumers. When we started 25 years ago, we had a difficult time convincing people that a beer that wasn't fizzy and yellow was something they should drink. That mystified me—why delicious beers like Long Trail were once a tough sell. Today, craft consumers are constantly looking for new and even extreme flavor profiles in the beers they buy. It's a complete turnaround in a generation, and I'm really in awe of today's craft fans. They constantly challenge and inspire us, and we love it.

LOST NATION BREWING

Making honest beer

Lost Nation Brewing crafts European-style ales and lagers inspired by the influences of Vermont. The beers and philosophy reflect founders Allen and Jamie's appreciation of quality and balance. They strive to expand people's palates by pushing the boundaries and limits of beer. Looking toward the future, Lost Nation intends to use this vision to produce beers that are wild, spontaneously fermented and barrel-aged. These beers will only be available in limited bottle-conditioned releases just a few times a year, so stop by the brewery frequently so you don't miss out!

Annual Production: 2,000 barrels

Fun Fact: The name Lost Nation comes more from folklore than fact—it is the land between two towns that is not maintained, incorporated or occupied. It is a place with no established norm, no preset rules. It is a place for everybody.

Don't Forget: When traveling Vermont's beer corridor, stop in to fill up a growler! All Lost Nation's growlers are filled to order with their counter pressure growler-filling machine. While you wait, be sure to check out the rotating menu of delicious, locally inspired pub dishes.

See page 181 for info on the taproom.

254 Wilkins Street
Morrisville, Vermont

- - - - -

802.851.8041

- - - - -

www.lostnationbrewing.com

JAMIE GRIFFITH + ALLEN VAN ANDA || Lost Nation

Q&A

What led you into the world of brewing?
Allen: In high school, I went to my local brewery in Red Bank, New Jersey, looking for a job. I told them I'd work for free, so they put me to work. A couple of weeks later, they offered me a job.
Jamie: Allen led me into the world of brewing. I'd been working in the food industry for a long time, but he taught me brewing principles and how to write recipes.

What's inspired your recipes and what are you experimenting with today?
Allen: Our palates dictate what we want to drink, what we want it to taste like. What we look at is how beer has evolved over hundreds of years and try to make something that goes with what we like to eat—something we can give to our friends and feel good about.
Jamie: Some of our beers are actual styles; others are a blend of styles. We look at a traditional style and write the recipe that way. Then, if there are a couple of beers that we think would go together style-wise, we put those together to create our own specific style.

What do you think the future holds for you as a brewer and your beer as a brand?
Allen: We want a good work environment, a good family life and a good community. One of the things that I think is exciting is that high-quality products are coming from Vermont, and beer is growing here. We're trying to foster that beer culture, and bring about a new way of looking at beer and beer consumption.
Jamie: I want to have a place where I look forward to coming to work, and that's this place. We'd like to stay hands-on, that's a big part of it—we always want to be making the beer.

> "We'd like to stay hands-on... we always want to be making the beer."

Your ideal day (involving beer of course!) . . .
Allen: Sitting on a back patio, eating oysters, enjoying a moment with my family.
Jamie: Just to have a minute to sit down and enjoy a beer with friends and family.

What intrigues, inspires or mystifies you?
Allen: One of the things I like most about beer is that there are so many variables. Yeast is a living organism, so it's always changing. If you're not learning something new, you're not paying attention to what's going on. Every day is different . . . that's enough to keep me interested all the time.
Jamie: The fact that every day I learn something.

MADISON BREWING CO.

Fermentation + civilization are inseparable

The Madison family has strong ties to their community—they've been in the Bennington area for almost 100 years! In the early 1990s, the Madisons decided to purchase what was then a burned-out storefront in Bennington's historic downtown and convert it into the area's first brewpub. In the fall of 1995, the Madison boys completed a brewer's training at the home of Shipyard Brewing Company then opened their pub and brewery in February of 1996. The Peter Austin seven-barrel brewing system is fired up approximately eight times a month by Head Brewer Michael Madison. He specializes in creating English ales and experiments with other styles as well.

Annual Production: 450 barrels

Fun Fact: Madison Brewing is located in what was a men's clothing store for more than 60 years called Adam's Clothing Store.

Don't Forget: The pub serves lunch and dinner and offers 10 handcrafted beers on tap—eight of their own brews as well as two seasonal offerings.

See page 183 for info on the pub.

BENNINGTON, VT

428 Main Street
Bennington, Vermont

- - - - -

802.442.7397

- - - - -

www.madisonbrewingco.com

Q&A

What led you into the world of brewing?

We entered the world of brewing blind. Our family has a long history of restaurant experience, but making beer was a new venture. After our purchase of a Peter Austin seven-barrel system, we headed to Federal Jack's in Kennebunk, Maine, the birthplace of Shipyard Brewing Company, to complete a two-week brewer's training.

What's inspired your recipes and what are you experimenting with today?

Originally our recipes were suggestions made by master brewer Alan Pugsley. With almost two decades of brewing experience behind us, we have certainly developed a style to our beers. One of our well-known seasonal beers that is appropriate to our location here in Vermont is the maple porter. We are also known for making a beer and then naming it after one of our customers. Our Wassicks White Wall Belgian-style beer was inspired by a local tire business.

What do you think the future holds for you as a brewer and your beer as a brand?

The increase in business at our restaurant has created the need to brew more frequently, and we have added four more beers to our taps to accommodate our growing customer demand. This has also brought about the desire to create new beers to provide a diverse selection for our customers' range of tastes.

> " I feel inspired to just create flavorful beer that my customers come back over and over again to drink. "

Your ideal day (involving beer of course!)

Taking a brewery tour of New England and trying out what the brewpubs have to offer for food and beverage. I enjoy the atmosphere of a brewpub and find myself looking to stop in and have a sample or two when the opportunity presents itself.

What intrigues, inspires or mystifies you?

I have always enjoyed an IPA but have never limited my brewing or my tasting to one style of beer. I feel inspired to just create flavorful beer that my customers come back over and over again to drink.

MAGIC HAT BREWING COMPANY

A performance in every bottle

Magic Hat was founded by Alan Newman and Bob Johnson in 1994. It was one of the first craft breweries in the state, and an early player in the craft beer scene nationwide. The first batch of Magic Hat was released to the public at the Burlington Winter Blues Festival. The beer was so well received, the entrepreneur-brewer duo decided to set up a brewery in a funky little space in Burlington. Magic Hat continues the tradition today, led by Head Brewer Chris Rockwood, of brewing distinctly flavored beers made with reference to ancient brewing traditions while adhering to high standards of quality, drinkability and pleasing their loyal customers.

Fun Fact: Magic Hat hosts the largest Mardi Gras Parade east of the Mississippi! Their annual event is a weekend-long celebration of music and the arts with show-stopping floats and, of course, great beer! Proceeds from the event benefit a local charity.

Don't Forget: Be sure to visit the Magic Hat Artifactory (pictured at left), where you'll usually find eight of their beers on tap, including year-round and seasonal offerings as well as some experimental brews.

5 Bartlett Bay Road
South Burlington, Vermont

- - - - -

802.658.2739

- - - - -

www.magichat.net

LEARN MORE ON PAPERPLATE.COM

CHRIS ROCKWOOD | *Magic Hat Brewing Company*

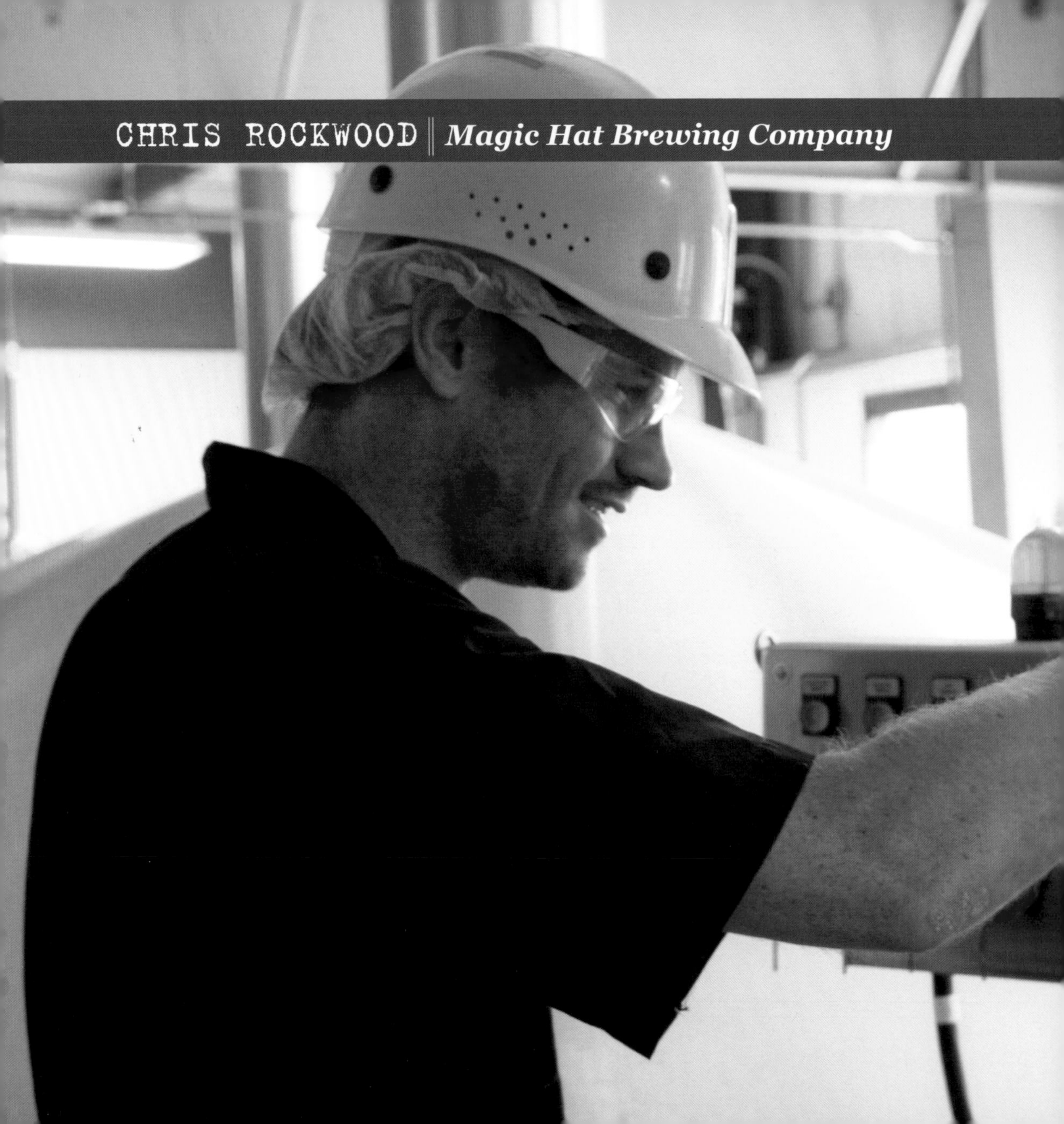

Q&A

What led you into the world of brewing?
I found my way into the brewing world through sampling. I grew up in a suburban town that did not offer a lot of variety in the world of beer. In college, I entered the beer section of a liquor store in Winooski and was filled with disbelief. I had no idea so many flavor combinations existed, but I knew I had to try them. This newfound curiosity, nourished by some time working as a sales rep in the beer industry and combined with a love of working with my hands, landed me where I am today.

What's inspired your recipes and what are you experimenting with today?
I strive to put forth flavor profiles that are well balanced, accessible and, most of all, intriguing. There are so many options available today, yet there are also a lot of new spaces to explore. Finding these spaces and blurring lines between styles is a really fun place to experiment. Barrel aging is a great way to accentuate flavors in any beer. What if we were to recreate the results of barrel aging using raw ingredients?

What do you think the future holds for you as a brewer and your beer as a brand?
As a brewer, there's always something to learn—whether it be a new technique, working with new materials or implementing old techniques in new ways. The future has untold possibilities and I look forward to refining the skills I have and acquiring as many new skills as possible. Working at a place with a history as rich and storied as Magic Hat's can be daunting at times. The road I now travel was paved by some really talented brewers who were willing to try some pretty crazy recipes. Finding ways to intrigue, excite and hopefully impress all those people is a challenge worth embracing.

> "I strive to put forth flavor profiles that are... intriguing."

Your ideal day (involving beer of course!) . . .
Beer has the unique talent of being able to enhance most days. No matter what your day has in store, there's a beer out there that can make it a bit more special. If you haven't found it yet, keep looking—someone is probably brewing it as we speak.

What intrigues, inspires or mystifies you?
For being something as simple as four ingredients, beer can be rather complex, straddling the line between science and art. One minute you are talking about adjusting a target number and the next you are lost in a conversation about subtle flavors of sweet, bitter, piney, citrusy or any other flavor you can imagine.

McNEILL'S PUB & BREWERY

Beer is the reason we wake up in the afternoon

Ray McNeill started brewing in 1991 and opened his brewpub in downtown Brattleboro that same year. He was one of the first four breweries in Vermont. To date, his beers have garnered 15 national and international awards at venues such as the Great American Beer Festival, the World Beer Cup, the United States Beer Tasting Championship and the Great International Beer Festival. Ray is focused on brewing American, British and German ales and lagers.

Fun Fact: Ray's first career was as a concert cellist. He studied music at Bennington College and CUNY Queens.

Don't Forget: McNeill's makes "beer that rocks and will enhance your life!" so fill up a growler on your next visit to the pub to see for yourself.

See page 185 for info on the pub.

90 Elliot Street
Brattleboro, Vermont

- - - - -

802.254.2553

- - - - -

www.facebook.com/McNeillsBrewery

LEARN MORE ON FARMPLATE.COM

Q&A

What led you into the world of brewing?
I initially owned a bar that sold international beers and very few American beers. I apprenticed at the long defunct Catamount brewery in the early 1990s, with the intention of adding a small brewery to the pub. Once I had begun, I was completely fascinated by both the art and science of the brewing process. (I am, at heart, a scientist and both of my daughters are scientists.) Fast-forward through several thousand pages of technical literature, and here I am today.

What's inspired your recipes and what are you experimenting with today?
First of all, I dislike that "recipe" word because I think it gives a very misleading snapshot of just how complicated my world really is. But, generally, my beers are inspired by traditional beer styles, coupled with what I actually enjoy drinking. I've never been able to make a Belgian tripel, for instance, because I really dislike them. The products I do make are all designed to emulate the traditional beer styles I enjoy. For instance, were I to try to make a Bavarian pilsner, I would begin by researching the typical pilsners of Bavaria, what malt they used, hops, gravity, bitterness, yeast, etc. And, with any luck, try to find some fresh samples (often difficult). As for modern American types, I certainly feel less constricted, and I've even created a couple for which I know of no prototype.

What do you think the future holds for you as a brewer and your beer as a brand?
As a brewer it's somewhat complicated because I'm a little disabled physically, but in my perfect world I would open a second brewpub. As for the brand? Who knows . . . I've already had beers in places I never foresaw. It's funny—20 years ago the *Financial Times* included us in an article about American beers to look out for, and I started getting all these phone calls from distributors and importers in England. At the time I was making 200 barrels a year and doing 100 percent of the work myself.

> " My beers are inspired by traditional beer styles. "

Your ideal day (involving beer of course!) . . .
I used to love mashing in. For me, it was nearly orgasmic. But I can't do that right now, so I don't know how to answer that. The ideal day ends, of course, with drinking a lot of beer.

What intrigues, inspires or mystifies you?
Yeast—fascinating critters. The way that malt, hops, pH and yeast all come together to form the perfect growing medium for brewing yeast is just mind-boggling.

NORTHSHIRE BREWERY

Handcrafted, classically styled session beers

Northshire Brewery was created by two friends who love beer and who love brewing it even more! Northshire Brewery makes classic beer styles but with their own twist. Beer's wide parameters allow founders Chris and Earl the freedom to use different ingredients and techniques to show off their individuality.

Annual Production: 1,000 barrels

Fun Fact: People are always surprised to see the amount of equipment packed into the small production room at the brewery. Given the confined space, brewing and packaging need to be carefully orchestrated.

Don't Forget: Northshire Brewery has a retail space where you can taste and buy their beer along with other merchandise. Cash sales only.

108 County Street
Bennington, Vermont

- - - - -

802.681.0201

- - - - -

www.northshirebrewery.com

What led you into the world of brewing?
Ever since I was old enough to drink, I have had a thirst for knowledge about how different styles of beer are made and where these different styles originate from. I've also always enjoyed classic styled beers, and I've always wanted to make my own versions with my own influence. Now I am!

What's inspired your recipes and what are you experimenting with today?
I am enjoying experimenting with cask conditioning beers, and I like to use authentic ingredients when available. I recently made a double IPA called Sicilian Pale Ale because it is made with Sicilian blood oranges and ancient Emmer grains. I love the process of developing recipes. I believe beer is the ultimate blank canvas.

What do you think the future holds for you as a brewer and your beer as a brand?
Hopefully we will continue to see growth in sales, and we will keep exploring different ingredients and styles as well as making our traditional year-round beers. We have plans to expand our tasting room and to continue to build our in-house lab to allow further testing and consistency in our beers.

Your ideal day (involving beer of course!) . . .
My ideal day would be to sit back with a cold glass of Northshire Chocolate Stout and watch as others lug heavy bags of grain and clean equipment in the 90°F brewery. Being able to take time to enjoy the beer that takes so much work to produce would be great.

"I love the process of developing recipes. I believe beer is the ultimate blank canvas."

What intrigues, inspires or mystifies you?
Yeast intrigues me in how it can impart so much character into a beer. I am always looking for more information about how my beers are developing. I am exploring the different components that affect beer, like water quality and yeast strains. The entire process of brewing is intriguing.

LEARN MORE AT FARMPLATE.COM

OTTER CREEK BREWING

One of Vermont's first craft breweries

Otter Creek is a proud leader in the American craft brewing revolution and has been producing award-winning beers for more than 20 years. The original brewery, where the first batch of Copper Ale was produced, is located just down the road from the current state-of-the-art facility, which has a 70,000-barrel capacity. All their beer is brewed in small batches using Vermont water, three strains of yeast and the finest hops and malts to ensure freshness and consistent quality. With their focus on sustainability, they are mindful of their contribution to the waste stream—their spent hops and yeast are donated to local farmers for use as an alternative fertilizer, and spent grains feed local dairy cows. You don't have to compromise your passion for great beer to promote environmental practices," says Brewmaster Mike Gerhart, "We prove that every day here."

Annual Production: 35,000 barrels

Fun Fact: Brewmaster Mike Gerhart started brewing beer when he was only 14 years old!

Don't Forget: The brewpub has large windows into the brewery, so you can see the beers you're enjoying actually being made!

See page 187 for info on the brewpub.

793 Exchange Street
Middlebury, Vermont

- - - - -

802.388.0727

- - - - -

www.facebook.com/OtterCreekBrewing

LEARN MORE ON PAMPLATE.COM

THE SHED BREWERY

Brewing handcrafted English-style ales

The Shed has been a Vermont staple for nearly 50 years. The original building in Stowe where the restaurant and brewery was housed was built in 1830 as a blacksmith shop. It later served as a cider mill and gathering spot for local farmers. During the harvest, farmers would drop off their apples and stay for a drink of hard cider and spirited conversation. As one of the first brewpubs in Vermont, the Shed's tradition of brewing handcrafted English-style ales remained part of Stowe's culture until 2011 when production shifted to Middlebury. Now, for the first time in its storied history, people outside of Vermont are able to enjoy Shed Mountain Ale. The legacy of the original lives on in every batch.

Annual Production: 12,000 barrels

Fun Fact: Along with the Mountain Ale recipe, Shed brewer Jim Lomax made the move from Stowe to Middlebury, where the Shed legend continues.

Don't Forget: The Shed was one of the first brewpubs in the East to offer beer to go in growlers. Don't forget to pick one up when you visit Otter Creek Brewing!

793 Exchange Street
Middlebury, Vermont

- - - - -

802.388.0727

- - - - -

www.facebook.com/theshedbrewery

WOLAVER'S FINE ORGANIC ALES

Better beer, better world

Nestled in the Green Mountains of Vermont, Wolaver's became the country's first USDA certified organic brewery in 1997. From Vermont-grown pumpkins and organic Vermont wildflower honey to locally roasted, fair-trade coffee, Wolaver's uses only the finest ingredients in their environmentally and socially responsible brewing processes to produce organic ales of the highest quality.

Annual Production: 8,500 barrels

Fun Fact: The pumpkins in their organic pumpkin ale are grown by Will and Judy Stevens of Golden Russet Farm in Shoreham, Vermont.

Don't Forget: Wolaver's is crafted at the Otter Creek facility and so gets its share of the talents of Brewmaster Mike Gerhart.

793 Exchange Street
Middlebury, Vermont

- - - -

802.388.0727

- - - -

www.www.facebook.com/wolavers

LEARN MORE ON FARMPLATE.COM

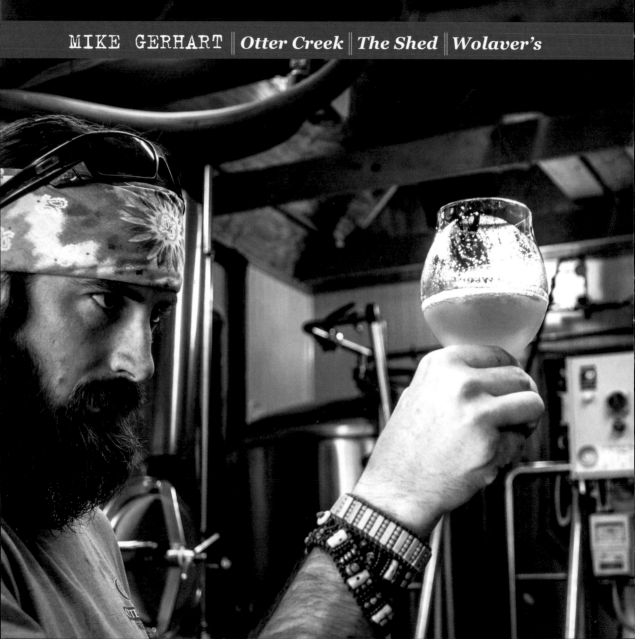

Q&A

What led you into the world of brewing?

I stumbled into the world of brewing a bit younger than most! I was a rambunctious kid and counselors suggested that my parents find me a hobby or prepare for the worst. Luckily they got me a homebrew kit, but it came with some ground rules: The beer I brewed had to remain at home and be consumed only with my parents present; I also had to take detailed tasting notes for each batch. So at just 14 years old, I was discussing malt profiles and hop flavors with my dad while my buddies were off getting into trouble. In 1996, I took my new craft and passion and headed to Vermont, where I brewed my way through college and beyond.

What's inspired your recipes and what are you experimenting with today?

I have beer and brewing on my mind pretty much all the time, so I tend to find inspiration for beers everywhere. It's not something I can turn off, and I end up with lots of journal entries about new combinations and pairings that look like the work of a madman—but that's my process. We recently purchased a mobile 15-gallon pilot brewing system that the rest of the brew crew and I work on all hours of the day and night. We seek out new hop varietals, vegetables, fresh produce—you name it. Everything on the table is worth trying!

What do you think the future holds for you as a brewer and your beer as a brand?

Otter Creek is one of New England's craft beer pioneers, and we continue to grow and strengthen with the experience that each new year brings. We believe in innovation and holding true to our core values, while also trying to stay ahead of the curve. As long as there are folks that demand more from their pint of beer, Otter Creek and I will be there to brew it!

> "I tend to find inspiration for beers everywhere."

Your ideal day (involving beer of course!) . . .

Some funky Hunter S. Thompson fusion of a VW bus, killer powder, the smell of steam rising from the whirlpool after the final hop addition and a bottomless keg. Throw in my pups and my best gal and you've got the stuff dreams are made of.

What intrigues, inspires or mystifies you?

Hands down, I am intrigued, mystified and inspired every day by the folks I get to work with. Everyone at OCB has a passion for beer that is shockingly contagious and drives us all to continually push the envelope in search of the perfect beer.

ROCK ART BREWERY

Brewing unique + traditional beers with local ingredients

Rock Art Brewery was born in 1997 from a home-brewer's desire to create fun, flavorful beers. In the early days, the brewery was located in the basement of Matt and Renée Nadeau's home. In 2011, the brewery moved into a state-of-the-art facility designed by Matt, which he describes as "just short of nirvana."

Annual Production: 3,500 barrels

Fun Fact: Rock Art uses a mixture of equipment to brew beer—from state-of-the-art lab apparatus to previously used and loved equipment, such as the eight tanks from the original basement brewery—whatever it takes to brew the best beer.

Don't Forget: Rock Art offers special releases throughout the year—from single-batch brews to a rotating hop IPA series, which showcases the flavors and aromas of the many new hop breeds available.

632 LaPorte Road (Route 100)
Morrisville, Vermont

- - - - -

802.888.9400

- - - - -

www.rockartbrewery.com

Q&A

What led you into the world of brewing?
I developed a taste for full-flavored beer at an early age, and I was always the odd duck who would order some obscure beer. Later in life, while living in Colorado, I had the pleasure of watching the Breckenridge Brewery & Pub being built. I could not wait to brew after that! And so began my homebrewing adventure. My love for cooking and a solid grasp of the sciences, as well as a talent for all things mechanical, paved the way for my new path.

What's inspired your recipes and what are you experimenting with today?
My recipes are born from flavors I want to try. I recall the process of creating our flagship beer Ridge Runner, a mild barley wine. I ordered up various grains and hops I hadn't brewed with before. I opened them to smell and taste them, just as a chef tastes ingredients. The two flavors that really grabbed my interest were the dark crystal malt, a very rich caramel toasty flavor, and the Challenger hop, which had an intense spicy, earthy, floral aroma. And so Ridge Runner was born. Currently we are experimenting with fresh pine tips that I picked in the spring, as well as preparing to brew our Vermont Hop Harvest Ale. We also started a rotating IPA series featuring either a single variety of hop or a blend of new varieties of hops.

What do you think the future holds for you as a brewer and your beer as a brand?
I look forward to brewing with more local ingredients and herbs. I have a few concepts in my head now—I just need to find an opportunity to brew them with our busy daily schedule. To all the wonderful hop growers out there I'd like to say, "Keep the new breeds coming—I'm having a ton of fun with them!"

> "I look forward to brewing with more local ingredients."

Your ideal day (involving beer of course!) . . .
My day is complete when I know we have brewed the best beer possible, I head home to spend time with our two boys and my wife Renée. I like to unwind in the evening with a beer or two, talking about the day or the future and what new adventures we might have as we round the next corner in the road of life!

What intrigues, inspires or mystifies you?
What is really mystical and intriguing is that, as brewers, we are creating flavorful sugar water. I enjoy looking at our fermenters and listening to the bubbles of carbon dioxide being released through the air lock as I visualize the beer inside the tank, with the yeast churning away as it creates alcohol. That is amazing!

Q&A

What led you into the world of brewing?
I began homebrewing because I couldn't find the beer I wanted in bottles. I discovered that I love the mixture of chemistry, biology and engineering involved in brewing, and it became a big part of my life. In part, we started Stone Corral for the same reasons. I love to brew and share beer. There's only so much double IPA you can drink. We figured if we felt that way, others might too.

What's inspired your recipes and what are you experimenting with today?
Our focus is on balance and flavor. I really enjoy pairing beer with food, so many of our recipes are structured to complement food. We brew variations of classic, well-balanced, sometimes hard-to-find styles. I couldn't find a good, brewery-fresh, traditional dunkelweizen locally, so that's one of the first styles we brewed. Our blonde has been a big hit, partly because it goes so well with a meal. Most of the blondes available in bottles are hopped clear into IPA territory. You can't really taste the subtlety of a fresh salad or a homemade marinade if you've seared your palate with a monster IPA or a quadruple stout. I like brewing hard-to-find traditional styles and letting people explore them. We'll be brewing some Belgian ales using very traditional hopping rates, some wild yeast, some local fruit. There is so much great beer that is still underrepresented in the brewing scene.

What do you think the future holds for you as a brewer and your beer as a brand?
We want to build on the balance displayed in traditional brewing styles and mix in our own flair and local flavors to bring truly fresh, local beer to our customers. Some say you can't really get a sense of *terroir* from beer, but you can work with the local water chemistry and create recipes to complement local flavors. We are a family-owned, artisanal brewery making distinctive styles, honoring some very old traditions, yet developing our own sense of place and community.

> "I like brewing hard-to-find traditional styles."

Your ideal day (involving beer of course!) . . .
Melissa's huevos rancheros for breakfast, work a half day in the brewery, take the afternoon off to swim, work horses or ski with family, enjoy a pint on the porch, grill some food, enjoy another pint . . .

What intrigues, inspires or mystifies you?
I get a sense of wonder from the array of flavors you can get with a handful of ingredients. I still remember the first time I tasted Chimay. It was unlike anything I had ever tasted. That you can take one wort and ferment it with different yeasts to end up with completely different beers still blows me away.

LEARN MORE AT FARMPLATE.COM

SWITCHBACK BREWING COMPANY

Brewing unfiltered ales since 2002

Switchback Brewing Company was founded in 2002 by Bill Cherry, a formally trained brewmaster. The vision was to concentrate limited resources into making just a few beers to achieve the highest quality possible. The success of the flagship beer, Switchback Ale, has driven constant expansion for more than 10 years and made the release of three rotating specials—Roasted Red Ale, Porter and Slow-Fermented Brown Ale—possible. As in the beginning, Switchback continues to rely on word-of-mouth to increase sales, which have expanded into New York, New Hampshire and Maine.

Fun Fact: The copper brewhouse used to brew all the Switchback ales was used for five decades at a brewery in Beerfelden, Germany, before making its way to Vermont in late 2008.

Don't Forget: Free tours are conducted every Saturday at 1 and 2 p.m. by reservation.

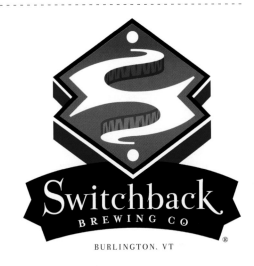

Switchback
BREWING CO

BURLINGTON, VT

160 Flynn Avenue
Burlington, Vermont

- - - - -

802.651.4114

- - - - -

www.switchbackvt.com

LEARN MORE ON FARMPLATE.COM

Q&A

What led you into the world of brewing?
Beer. And the process of making beer. At its best, brewing is a wonderfully complex manipulation of completely natural processes, with a result different from nature's original intention.

What's inspired your recipes and what are you experimenting with today?
I am inspired by the greatest beers made in the world and the experimentation of what the underlying principles are that make them so good and timeless. My goal is always to build from these ideas something unique that adds to brewing culture.

What do you think the future holds for you as a brewer and your beer as a brand?
I am very excited for the future of craft brewing. We have extensive plans to increase our technical capabilities so we can experiment with difficult new brewing techniques as well as forever improving the quality and consistency of our beers. We take an approach toward brewing whereby the brewer's expertise and skill is always at the center of the craft, meaning in part that we choose not to automate the brewing process, even though we do invest in high-tech equipment.

Your ideal day (involving beer of course!) . . .
My ideal day is brewing day! I love everything about making beer, including insanely cleaning everything. There is nothing better than working nonstop all day without a break, and seeing that full tank of beer at the end of the day. It is my definition of a good kind of tired.

> Brewing is a wonderfully complex manipulation of completely natural processes..."

What intrigues, inspires or mystifies you?
The subtlety of flavor combinations. In brewing we are getting flavors from water, malt, grains, hops and, most importantly, yeast. What mystifies me is how sometimes individual flavor components can be downright awful tasting, yet when properly combined in harmony, they can be the crucial components that make the overall flavor experience transcendent and wonderful.

TRAPP LAGER BREWERY

Austrian-inspired lagers

More than a decade ago, Johannes von Trapp started thinking about brewing his own beer for guests of the Trapp Family Lodge. His dream was to produce an American version of the lager he enjoys on his trips to the countryside near his ancestral home in Austria. His dream became a reality in the spring of 2010 with the opening of the Trapp Lager Brewery. The current brewery, situated on the lower level of what was once the Austrian Tea Room, is modest in size and production capabilities, but construction is well underway on a brand new 40,000-square-foot brewery including a 5,000-square-foot beer hall, scheduled to open in the summer of 2014.

Annual Production: 1,875 barrels

Fun Fact: The source of pure spring water for Trapp lager is an artesian well near the brewery. The spring water has chemical qualities similar to Austrian spring water and is considered perfect for brewing European-style lagers.

Don't Forget: The Trapp Lager Brewery is located on a 2,500-acre mountaintop resort owned and operated by the von Trapp family, whose story inspired *The Sound of Music*.

700 Trapp Hill Road
Stowe, Vermont

- - - - -

802.253.8511

- - - - -

www.trappfamily.com/amenities/dining/brewery

Q&A

What led you into the world of brewing?

My career began at Plymouth State College, New Hampshire, in a dark and dusty basement with my three roommates as taste testers. After graduation, all my friends got jobs in cities, but I packed up and headed for Burlington, where I got a dream job as a brewer for Magic Hat. Flash-forward a decade and I found myself the general manager of what was at the time the seventh largest craft brewery in the country. The irony was, I was commuting 40 minutes to work, and I was working in a city! Times change, and I started at Trapp Lager Brewery in October 2012. No more waiting in traffic—now I have a breathtaking view of the Worcester Range right from the brewery, and the only lines are when I'm waiting to get on the gondola!

What's inspired your recipes and what are you experimenting with today?

All our recipes are hundreds of years old and have been produced many times by many different brewers. Our styles have my imprint on them and are my interpretations of how a true Austrian-style lager should be. I've been tinkering with some true Austrian/German styles in my garage recently. I've been trying to work out some kinks on a Hefeweizen and dark bock. My next recipe will be in the direction of a shandy-style German beer.

What do you think the future holds for you as a brewer and your beer as a brand?

We are living in a trendy world—originality has become a trait that is truly hard to find. We are experiencing a boom in hoppy beers today. Many brewers are making outrageous pallet-busting beers that appeal to the uber beer geek. Sometimes the profile is so hoppy it hides a lot of imperfections. We make clean Austrian beer. No gimmicks, no bells and whistles and no funky names. We are going to bring craft lager to the masses, a true representation of the style. Very straightforward.

> " We are going to bring craft lager to the masses... "

Your ideal day (involving beer of course!) . . .

Dueling banjos, an ice-cold Vienna lager, hanging with my friends and family around a fire at the end of an epic powder day—that is living.

What intrigues, inspires or mystifies you?

What inspires me? Evolution baby. What intrigues me? Constantly getting better as a brewer, father, husband and overall person. And mystifies me? This one definitely has to be the ocean—scary place out there.

TROUT RIVER BREWING COMPANY

Catch some trout

Trout River Brewing was founded in July 1996 by Dan and Laura Gates. They began brewing in early December of that year. Dan still owns and operates this microbrewery in the Northeast Kingdom of Vermont. Trout River Brewing prides itself on creating and serving fresh, all-natural premium ales and lagers. Their beers are made with pure Vermont water, the best imported and domestic malts and hops and quality yeasts. Nothing else is added and nothing is taken away!

Annual Production: 3,000 barrels

Fun Fact: Trout River was one of the first micro-breweries in Vermont—only Catamount, Long Trail, McNeill's, Otter Creek and The Vermont Pub & Brewery were on the scene when Trout River was founded.

Don't Forget: Their pub serves gourmet, hand-tossed sourdough pizzas as well as their own ales and lagers on tap every Friday and Saturday evening from 4 to 9 p.m.

See page 202 for info on the pub.

Brewed in Vermont • Naturally Unfiltered

645 Broad Street
Lyndonville, Vermont

- - - - -

802.626.9396

- - - - -

www.troutriverbrewing.com

DAN GATES | *Trout River Brewing Company*

Q&A

What led you into the world of brewing?
I was into cooking, and it went along with the cooking thing because I was always trying beverages and foods. And I used to make hard cider when I was young, growing up on the farm, so picking up the brewing thing was a natural, easy progression.

What's inspired your recipes and what are you experimenting with today?
My main signature thing that I try to do with all my beers is make them smooth, clean finishing and well balanced so they are really drinkable. I even have some really strong beers that are very drinkable—like the Knightslayer, which is our imperial stout, and the Boneyard Barleywine, which I started making back when I was homebrewing. As for new, we are going to do some Belgian beers. I really do enjoy Belgian beers.

What do you think the future holds for you as a brewer and your beer as a brand?
I know if I wanted to go out and do it, the brand could be built because it hasn't had that much exposure as far as out of state. But that much marketing isn't my cup of tea. Ideally, from my standpoint, someone would buy the brand because, at this point in my life, I'd like to step back from the crazy high-paced marketing part and just be a pub and brewery.

Your ideal day (involving beer of course!) . . .
Watching football on Sunday and drinking beer. And cooking—I'd probably do ribs or a smoked pork butt, and I'd probably have some bourbon or scotch in there too since it's a football game. From a production standpoint, brew day is nice, but only on a cool day. You come in, it's physical, it's steamy. Or pub night on weekends when people are all excited to taste the beer that's in stock.

What intrigues, inspires or mystifies you?
Good food and beverage is always an inspiration, whether it's mine or someone else's. What intrigues? Pretty much getting out on my mountain bike or skiing are things that I want to do when I get out of the brewery, so that's what I'm thinking about. Mystified? Oh my God. That'd be—how the hell did I get into this stuff? It's been 17 years, and on days when things go bad, I just wonder, "Why?"

> " Good food and beverage is always an inspiration... "

THE VERMONT PUB & BREWERY

Pioneers of Vermont craft beer

Craft beer pioneer Greg Noonan opened The Vermont Pub & Brewery in downtown Burlington in 1988. Vermont's original brewpub introduced a variety of freshly brewed beers to its patrons, including the renowned Burly Irish Ale, which is still on tap! Today, fueled by the core values of quality food and beer, knowledge-able service, affordable prices and a family-friendly atmosphere, VPB enjoys continued success and remains an anchor in the craft beer segment. Under the direction of 14-year veteran brewmaster Russ FitzPatrick, VPB will continue to push the envelope with creative and innovative beer, incubated in the exper-imental brewery, while offering compatible food choices to enhance the pub experience.

Annual Production: 1,000 barrels

Fun Fact: Greg Noonan single-handedly designed and built the 15'x12' sign that hangs on the brewery's exterior. It took him seven months.

Don't Forget: The relaxed pub offers delicious, reasonably priced food and a variety of beer flights. The fresh beer made on premises never sees the light of day!

See page 205 for info on the pub.

144 College Street
Burlington, Vermont

- - - - -

802.865.0500

- - - - -

www.vermontbrewery.com

LEARN MORE ON FARMPLATE.COM

135

STEVE POLEWACYK || *The Vermont Pub & Brewery*

Q&A

What led you into the world of brewing?
I entered the craft brewing scene over 25 years ago, through the "back door" so to speak. After sharing our first brew together in 1974, Greg Noonan *[pictured with Steve at left]* and I had always enjoyed discussing the many curiously interesting European beers we had tried. Fast forward 15 years, and Greg was opening a brewpub. I came in to help with his start-up and ended up staying.

What's inspired your recipes and what are you experimenting with today?
Greg originated every one of our recipes until passing away in 2009. As early as 1989, he was creating Brettanomyces beers, spruce beers, fruit lambics and many other styles that were pushing craft beer forward. In 1988, there were 22 recognized beer styles; today, the Brewers Association touts over 100. Greg was a progenitor of that movement. We continue in his spirit by experimenting with such things as new yeast strains, 100 percent natural adjuncts and barrel aging beers. Several years ago, VPB installed a small pilot brewery within our original cramped brewhouse to continue Greg's pioneering efforts. Last year alone, we brewed more than 60 different styles of beer.

What do you think the future holds for you as a brewer and your beer as a brand?
Our beer as a brand is only a part of the larger picture. We have been honored through the years with over 50 awards, medals and ribbons for our brewing efforts. As we move forward, it is important to foster our original vision of good food and great beer served by a knowledgeable staff in a welcoming atmosphere. So far, so good. We have recently completed renovations to the pub, and we have made enhancements to our cellar brewery with the goal of increasing efficiency.

> "I'm inspired by those who've been impacted by Greg's influence."

Your ideal day (involving beer of course!) . . .
Tasting a variety of our finely crafted beers along with some appetizers in the company of friends. Discussing flavor profiles and simply talking about the nuances of these pairings always ends up with a bunch of happy people, sitting around enjoying life.

What intrigues, inspires or mystifies you?
I am truly intrigued by the evolution of craft beer both here and abroad. I've seen an incredible amount of progress over the last 25 years and I'm inspired by those who've been impacted by Greg's influence. His legacy is truly realized through the success of many renowned brewers of today.

LEARN MORE ON FARMPLATE.COM

WHETSTONE STATION

Craft brews + amazing views

Whetstone Station Restaurant and Brewery is a very small three-barrel brewery combined with a large casual restaurant located right on the Connecticut River. Along with several always-changing unique craft beers of their own, they offer inspired food, dozens of guest craft beers and specialty drinks—all in a beautiful waterfront location. Originally the building was a gashouse that supplied Brattleboro's street lamps. Its unique round shape has been restored into a beautiful space with a contemporary, industrial vibe. The restaurant's unusual location—perched above the river, alongside a steel bridge and nestled amongst the Green Mountains—makes it a special place to visit.

Annual Production: 900 barrels

Fun Fact: When Whetstone Station's brewery needed to expand their cellar size, the chiller needed to be replaced. In a nod to their use of reclaimed materials, they were able to source their new chiller from the recently retired MRI machine from the local hospital—likely making their craft beers the only ones chilled by medical equipment!

Don't Forget: There is both indoor and outdoor dining, and the restaurant is just a short stroll from downtown Brattleboro.

See page 206 for info on the restaurant.

36 Bridge Street
Brattleboro, Vermont

- - - - -

802.490.2354

- - - - -

www.whetstonestation.com

Q&A

What led you into the world of brewing?
I started homebrewing in 1996 in college, primarily at the time as a way to have a large supply of inexpensive beer. I get really into anything I am passionate about. So from there, it became a quest to make the best beer that I could in ways that were not always traditional.

What's inspired your recipes and what are you experimenting with today?
I am a big fan of variety. I am a larger brewery's nightmare customer because I am one of those guys that's always looking to try something new. When I finally decided to start a nano of my own, I made the promise to myself that I wouldn't make the same beer twice. So each time we brew here, we publish the recipe on our website so that if you really enjoyed it, you can make it yourself. Even when we repeat a style, or find a combination of grain or hops we love, with our next use we always tweak something. It probably stems from my being my own worst critic. As far as experimenting goes, we are working with someone right now on a house yeast strain, isolated from an old farm property we have just a few miles from the brewery.

> " It became a quest to make the best beer that I could in ways that were not always traditional. "

What do you think the future holds for you as a brewer and your beer as a brand?
We're super small, brewing just around 100 gallons per batch. We are planning to stay that way. Our future will include oddball styles and processes, trying new ways to get from point A to B. There's been so much exciting growth and innovation in the brewing world recently, we look forward to being a part of that.

Your ideal day (involving beer of course!) . . .
Well, an ideal non-work day would probably be a few good bottles of creative brews from some of my favorite brewers, on a beach somewhere . . .

What intrigues, inspires or mystifies you?
I love sours—love 'em. Ever since Chris Lively of Ebenezer's Pub in Lovell, Maine, introduced me to the world of funky sour beer in 2008, I have been hooked.

ZERO GRAVITY CRAFT BREWERY

Brewed for food

Zero Gravity Craft Brewery uses only the finest available ingredients and brews with painstaking attention to detail. The brewery focuses on some of the lesser known and more difficult-to-make beer styles traditionally found in Germany, Belgium and the UK. As often as possible, locally sourced ingredients are used in the brewing process. The brewery's location at American Flatbread provides a vehicle for high-lighting the food-friendly nature of many Zero Gravity beers.

Annual Production: 1,500 barrels

Fun Fact: A beer with zero gravity wouldn't be much of a beer at all! In brewing terms, the starting gravity of a beer is directly propor-tional to the amount of alcohol in the finished product. The name is a nod to the brewery's commitment to limiting its environmental footprint as well as how one might feel after enjoying a few!

Don't Forget: The brewery's home is at American Flatbread Burlington Hearth, but you can also find Zero Gravity at better restaurants and bars throughout Vermont.

See page 154 for info on the restaurant.

115 St. Paul Street
Burlington, Vermont

- - - - -

802.861.2999

- - - - -

www.zerogravitybeer.com

Q&A

with Paul Sayler

What led you into the world of brewing?
A couple of notable experiences influenced me going into beer as a profession. The first was returning to Seattle, my birthplace, in 1981 and drinking some Grant's Imperial Stout—it's an epiphany when you cross paths with something that has that degree of flavor and verve. The next was while visiting Germany—I was quietly blown away by the age-old traditions the Germans took for granted and the superior quality of their beers. After college, I realized I was really intrigued by an offer I'd received to apprentice at Catamount Brewing and so I decided to pursue the opportunity.

What's inspired your recipes and what are you experimenting with today?
My early career was focused on trying to master ancient styles as well as the traditional beers of Europe. My biggest passion right now, however, is pairing work—developing an understanding of how food works with beer. I'm very interested in treating the brewhouse as a kitchen and working to develop beers that are truly food-friendly. We just brewed a garden beer, with roasted squash and carrots, and we're also doing infusions of herbs and spices.

What do you think the future holds for you as a brewer and your beer as a brand?
When we were getting our restaurants to a place of success, brewing for me had to be put aside. I'd like to get back to focusing on beer 100 percent. It's pretty exciting to be reentering the scene I've been part of from the beginning at a moment when Vermont is becoming one of the new brewing meccas. Our brand is something that is evolving within that setting, and hopefully it will benefit from the perspective I've gained in the industry over time.

> "My biggest passion right now is...how food works with beer."

Your ideal day (involving beer of course!) . . .
After touring breweries that have a significant place in my imagination (in Hamburg, Prague, Edinburgh or Denver), spending the evening drinking that beer with a meal that is true to place.

What intrigues, inspires or mystifies you?
I'm inspired by my fellow brewers and what seems to be the wellspring of energy that keeps this movement growing at a rate I could never have imagined. I'm intrigued by the deep history of beer. As for mystifying, the power of that first taste of a newly fermented batch reminds me I'm greeting something I've never crossed paths with before. That sense of mystery is key to why I've stayed a brewer.

EAT & DRINK

EAT GOOD FOOD ... *drink great beer*

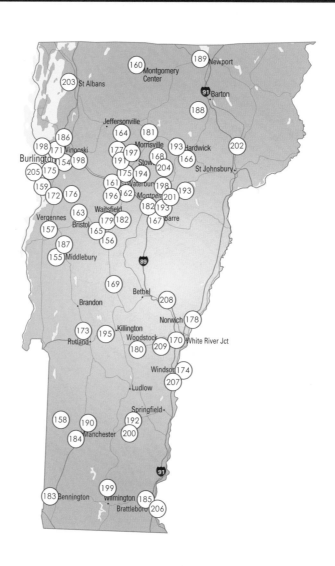

EAT & DRINK

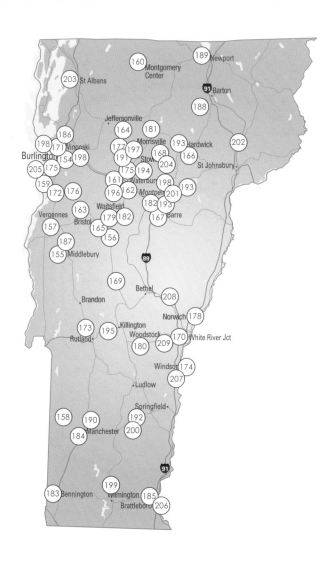

189 Newport

160 Montgomery Center

203 St Albans

91 Barton

188

Jeffersonville

186

171 Winooski

198

Burlington

154 198

205 175

159

172 176

164

181

177 197

191

175 194

161

196 162

Morrisville

168

204

Stowe

198

201

182 193

167

193 Hardwick

166

St Johnsbury

202

193

Waterbury

Montpelier

163

179 182

165

156

Waitsfield

182

Barre

Vergennes

157

187

155 Middlebury

169

Brandon

Bethel

208

Norwich 178

173

195

Killington

Woodstock

180

209

170 White River Jct

Rutland

Windsor 174

207

Ludlow

Springfield

158

190

184 Manchester

192

200

89

91

183 Bennington

199

185

Wilmington

206

Brattleboro

EAT & DRINK

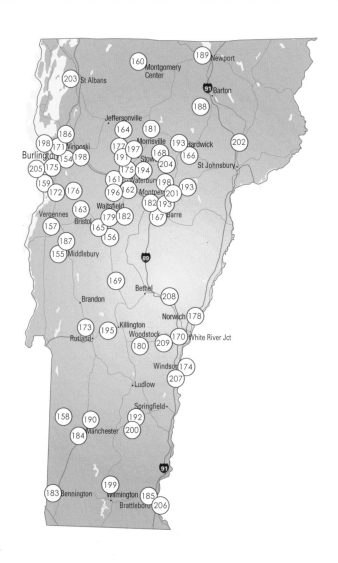

EAT & DRINK

AMERICAN FLATBREAD
BURLINGTON HEARTH

Wood-fired pizza + home to Zero Gravity Craft Brewery

Since opening in 2004, American Flatbread Burlington Hearth has played a central role in Burlington's craft beer renaissance. Zero Gravity Craft Brewery anchors the tap lineup, rotating upwards of 18 unique craft beers on draft and hand-pulled beer engines. The restaurant and brewery work closely together to create a beer and food pairing experience unlike any other in the area. Beer dinners are held most months throughout the year, bringing local ingredients together in several courses, each paired with a different Zero Gravity offering.

Burlington Hearth

Fun Fact: Some say the restaurant's location is haunted. To appease the spirits, an offering of Zero Gravity is routinely left on the bar at night.

Don't Forget: Don't forget to check out Saturday and Sunday Brunch where Eggs Benedict Flatbreads are served alongside house Bloody Marys and Zero Gravity beer backs.

See page 143 for info on the brewery.

115 St. Paul Street
Burlington, Vermont

- - - - -

802.861.2999

- - - - -

www.americanflatbread.com

AMERICAN FLATBREAD
MIDDLEBURY HEARTH

All-natural flatbread pizza

Experience fireside dining and made-to-order flatbreads fresh from their signature earthen oven. Along with salads, desserts, natural beverages, wines and microbrews, American Flatbread offers all-natural flatbread pizzas featuring ingredients raised in Vermont as well as right in Addison County that will please the pickiest two-year old, the hungriest college kid or the most experienced pizza connoisseur!

Fun Fact: The outdoor seating area features a fire circle to sip your beer by and meet new friends or just gaze into while waiting for your table.

Don't Forget: The space is great for large groups. Call or email ahead to make a reservation.

137 Maple Street
Middlebury, Vermont
- - - - -
802.388.3300
- - - - -
www.americanflatbread.com

AMERICAN FLATBREAD
WAITSFIELD HEARTH

All-natural pizza baked in a primitive oven

Lareau Farm is home to the original American Flatbread, founded in 1985 by social entrepreneur and community advocate George Schenk. American Flatbread is renowned for producing tender, crisp flatbreads from a wood-fired earthen oven. Their flatbreads are topped with unique combinations of the finest local and regional ingredients. The casual, rustic atmosphere, delicious farm-to-table-inspired menu and variety of local and other microbrews on tap make this a favorite destination for locals and visitors alike.

Fun Fact: The flatbread oven was designed and built by founder George Schenk himself. He constructed it with rock and clay collected near the farm.

Don't Forget: Lareau Farm is situated on the bucolic banks of the Mad River. In addition to the restaurant and 13 comfortable rooms where you can spend the night, Lareau Farm has swimming holes on the river as well as biking and walking trails!

46 Lareau Road
Waitsfield, Vermont

- - - - -

802.496.8856

- - - - -

www.americanflatbread.com

BAR ANTIDOTE

Restaurant + cocktail lounge

This welcoming restaurant in downtown Vergennes has great food, a friendly staff, a comfortable and stylish bar, beautiful local artwork on the walls, live music and entertainment, as well as indoor and seasonal outdoor seating. Bar Antidote is a craft-beer lover's destination, with a rotating selection of 14 craft beers on tap.

Fun Fact: All burgers are sourced within 13 miles of the restaurant, hence their 13-mile burger menu.

Don't Forget: Bar Antidote features a full lineup of live music every Wednesday and Saturday nights.

35 Green Street
Vergennes, Vermont

- - - - -

802.877.2555

- - - - -

www.barantidote.com

BARROWS HOUSE RESTAURANT

Traditional pub with a modern twist

Barrows House Restaurant serves traditional New England tavern fare with a modern twist, complemented by an extensive selection of craft beers. The menus and drinks showcase Vermont's abundance of local products and celebrate the state's farms and producers. These quality ingredients are simply prepared using classic culinary techniques to offer guests an uncomplicated seasonal taste of Vermont. Set against the historic backdrop of an 1803 building, the ambiance is at once warm and welcoming, sophisticated yet casual.

Fun Fact: The building features Dorset white marble accents and historic black-and-white photos from the local Norcross-West Marble Quarry, one of the oldest in the country.

Don't Forget: There is a large outdoor terrace with a wood-burning fire pit surrounded by colorful gardens, which makes it an ideal setting for outdoor dining and all types of special events.

3156 Route 30
Dorset, Vermont

- - - - -

802.867.4455

- - - - -

www.barrowshouse.com

THE BEARDED FROG

Eclectic American fare

At The Bearded Frog, the drink selections match the eclectic and unique nature of the dining menu. A comprehensive wine list is carefully selected from mostly small production vineyards around the world. The cocktails are all original and handcrafted using fresh herbs, fruits and juices whenever possible. Draft beer offerings are selected exclusively from Vermont and New England microbreweries and are complemented by a carefully chosen artisanal bottle list representing breweries around the world.

Fun Fact: Beards are encouraged, but not required.

Don't Forget: The Bearded Frog is only open for dinner, but try breakfast or lunch at the Next Door Cafe located right . . . next door!

5247 Shelburne Road
Shelburne, Vermont

- - - - -

802.985.9877

- - - - -

www.thebeardedfrog.com

THE BLACK LANTERN INN BREWPUB

A classic Irish-style pub serving handcrafted beer

The Black Lantern Inn has been in continuous operation since 1803. The dining room and brewpub feature a rotating assortment of local and regional microbrews on tap as well as their own handcrafted beer. Hearty tavern fare is served in both the Irish brewpub setting as well as in the quieter, romantic dining room, where patrons can enjoy daily food and whiskey specials.

Fun Fact: The Black Lantern Inn features the area's largest selection of Irish whiskey—36 and counting!

Don't Forget: The inn is the only place you'll find their beer. Make a night of it and plan to stay in one of their comfortable rooms!

See page 29 for info on the brewery.

2057 North Main Street
Montgomery, Vermont

- - - - -

802.326.3269

- - - - -

www.theblacklanterninn.com

BLACKBACK PUB

Craft beer pub + flyshop

Blackback usually runs 18 to 22 beers on tap. They are committed to selling only the very best beers and feature local craft breweries like Hill Farmstead, Lawson's Finest Liquids and Zero Gravity. They also serve exotic wild-yeast beers from all over the world, top session beers from Germany and the Czech Republic as well as offerings from international gypsy breweries like Mikkeller and Evil Twin. They also have a top-notch whisky selection.

Fun Fact: Blackback is a nickname for brook trout, the only stream-dwelling trout native to New England. Owner Rick Binet is a fly-fishing guide for salmon and trout in northern Vermont.

Don't Forget: The Mad Taco *(see page 182)* offers fresh Mexican fare from the Blackback kitchen!

1 Stowe Street
Waterbury, Vermont

- - - - -

802.244.0123

- - - - -

www.blackbackpub.com

THE BLUE STONE

Serious pizza. Humble food. No bull.

The Blue Stone is a rustic pizza shop and tavern in the heart of Vermont featuring old-world, hand-tossed pizza with a new local twist. They begin with fresh dough, bread and sauces made by hand every day. They add the best all-natural ingredients they can find and prepare many unique toppings in-house to create a new kind of pizza. As the seasons change, so do their offerings. They will always be working with whatever is fresh, local and in season to offer new farm-inspired variations on what pizza used to be. To complement their hand-crafted pizza, The Blue Stone features a casual pub menu and a rotating draft list of 19 craft beers from around Vermont and beyond.

Fun Fact: Their namesake rock—The Blue Stone—is a vintage well stone the owners resurrected from a circa 1700s farmhouse in Pittsford, Vermont. This relic lives on as the community bar table—the ideal spot to grab a slice and a beer!

Don't Forget: The Alchemist's Heady Topper in cans is available every week while it lasts—only five bucks.

15 Stowe Street
Waterbury, Vermont
- - - - -
802.882.8185
- - - - -
www.bluestonevt.com

BOBCAT CAFE & BREWERY

Contemporary comfort food + house-brewed beer

The Bobcat Café focuses on local ingredients to enhance their creative comfort food. Designed to be the "third place" (a social retreat from home and work), Chefs Erin and Sanderson change the menu with each season and strive to create food that is interesting and wholesome enough to ensure locals and visitors return often. The Bobcat has developed strong relationships with farmers, butchers and cheesemakers in Addison County in order to bring freshness and a sense of place to the table. And the beer couldn't be fresher—all their beers on tap are brewed in house!

Fun Fact: The Bobcat Café started as a community-supported restaurant back when it opened in 2002. To this day, the restaurant continues to be community minded by hosting weekly benefit dinners and donating 20 percent of food sales to local nonprofits or charities.

Don't Forget: Come to the bar between 4 and 5 p.m. to sample the Bar Bites menu, featuring a create-your-own platter with various snacks and local cheeses, the Bobcat's infamous Misty Knoll chicken wings as well as a daily $5 early-bird special.

See page 33 for info on the brewery.

5 Main Street
Bristol, Vermont

- - - - -

802.453.3311

- - - - -

www.bobcatcafe.com

BREWSTER RIVER PUB & BREWERY

Locally renowned for great food, entertainment + beer

Brewster River's goal is to provide their guests with the best possible experience by offering homemade foods made with the finest fresh, local ingredients and served by a friendly, knowledgeable staff. They support New England's vast selection of craft brewers with frequently rotating guest taps to complement their own craft brews, offering beers for all palates!

Fun Fact: The pub has live music several times a week, featuring Vermont's very talented musicians and bands.

Don't Forget: Brewster River is located at the base of Smugglers' Notch Resort and is a renowned après-ski destination.

See page 37 for info on the brewery.

4087 Route 108 South
Jeffersonville, Vermont

- - - - -

802.644.6366

- - - - -

www.brewsterriverpubnbrewery.com

CASTLEROCK PUB

A mountain pub at Sugarbush Resort

Castlerock Pub serves chef-inspired better-than-bar food along with a fine selection of the best Vermont craft brews, wines and spirits. Conveniently situated at the base of the slopes in the Gate House Lodge, the pub is open daily for lunch and après-whenever-the-lifts-are-spinning (both winter and summer). The pub also features live music on winter weekends.

Fun Fact: The Bernese mountain dog puppy featured on the pub's sign is Rumble, resort owner Win Smith's dog. Rumble's doghouse sits across the courtyard from the pub.

Don't Forget: Castlerock Pub hosts the Sugarbush Brew-Grass Festival in June with over 30 craft brewers serving their beers amidst live music by local bands. It's a quintessential way to kick off summer! The pub also has specialty beer dinners throughout the year featuring a craft brewer who carefully pairs a selection of beers with a locally inspired dinner.

Lincoln Peak at Sugarbush Resort
Sugarbush Access Road
Warren, Vermont
- - - - -
802.583.6594
- - - - -
www.sugarbush.com

LEARN MORE ON FARMPLATE.COM

165

CLAIRE'S RESTAURANT & BAR

Community supported restaurant

Claire's serves farm-to-table cuisine, emphasizing local and sustainable produce, artisanal products and responsible business practices that support our communities. Their cooking philosophy is inspired by flavors of the world. They have four taps usually featuring two Hill Farmstead beers, Citizen Cider and a rotation of Switchback, Lost Nation, Rock Art, Trout River, Wolaver's and Otter Creek, to name just a few. Claire's mission emphasizes supporting local producers of artisanal, sustainable products, and they are fortunate to be able to feature such amazing beers, brewed right in their backyard!

Fun Fact: A local dairy farmer picks up the used fryer oil from Claire's kitchen to use as biodiesel for his farm equipment.

Don't Forget: There's live music on Thursday nights and drink specials every Sunday and Monday.

41 South Main Street
Hardwick, Vermont

- - - - -

802.472.7053

- - - - -

www.clairesvt.com

CORNERSTONE PUB & KITCHEN

Revitalizing Barre one pint at a time

Established in 2012, Cornerstone Pub & Kitchen is a modern American gastropub with 28 draft beers and a diverse menu, including a secret recipe for their fish and chips batter made from a blend of draft beers. The pub is a collaboration between two lifelong best friends with over 40 years of restaurant experience between them. They have brought together their passions for the finest craft beers and good, honest food to create a unique gathering place in their hometown.

Fun Fact: Cornerstone isn't just a clever name for the grand stone building located on the corner of Main and Elm streets. It has deeper meaning to owners Keith and Rich—their goal is to be the cornerstone of their community and to help revitalize Barre "one pint at a time."

Don't Forget: While the pub only takes reservations for parties of five or more, smaller parties can call shortly before arriving and they will add you to their list.

CORNERSTONE
PUB & KITCHEN

47 North Main Street
Barre, Vermont
- - - - -
802.476.2121
- - - - -
www.cornerstonepk.com

LEARN MORE ON FARMPLATE.COM

CROP BISTRO & BREWERY

Share the bounty of Crop

Crop Bistro & Brewery serves simple flavorful fare inspired by the freshest local and regional ingredients. Along with great food, the bistro has 10 beers on tap, at least five of which are usually their own, as well as an extensive list of other local, national and international ales and lagers. The bistro has an interesting wine list as well as simple, innovative cocktails made from local hard cider and spirits. The beautiful interior spaces feature photographs by renowned local photographer Peter Miller, as well as a legendary pub and handcrafted bistro bar.

Fun Fact: The kitchen at Crop Bistro & Brewery is helmed by Tom Bivins, named 2011 Vermont Chef of the Year.

Don't Forget: Tours are offered on Saturdays at 4 p.m. or by appointment or chance. And be sure to try all their delicious beers on tap by ordering a sampler rack!

See page 41 for info on the brewery.

1859 Mountain Road
Stowe, Vermont

- - - - -

802.253.4765

- - - - -

www.cropvt.com

DOC'S TAVERN

A New England public house

Doc's Tavern is housed in a newly renovated barn adjacent to the The Huntington House Inn. The tavern is a favorite destination for both locals and visitors alike and has plenty to entertain—a pool table, shuffleboard, music and several large screen TVs. Relax at the bar or in one of the cozy booths and enjoy traditional pub food and drink. The tavern has 12 taps, most of which feature Vermont's finest brews.

Fun Fact: Doc's Tavern proudly serves local beef.

Don't Forget: Located in the very heart of Vermont, The Huntington House Inn can be your headquarters for exploring Vermont in every season. While it's close to several of the areas finest ski resorts and tourist attractions, it's far enough away to avoid the madding crowds.

19 Huntington Place
Rochester, Vermont

- - - - -

802.767.9140

- - - - -

www.huntingtonhouseinn.com

Elixir's menu features creative, approach-able food that is freshly prepared using local ingredients. All of their draft beers are sourced from Vermont and the selection constantly rotates. they also have an innovative cocktail list, and the interesting and affordable wines have been carefully selected to complement the menu. Whether you simply want to relax with a glass of wine or enjoy dinner from the ever-changing seasonal menu, Elixir offers a warm, inviting atmosphere with friendly, knowl-edgeable service.

Fun Fact: As with everything else, the ice creams are made in-house daily using only the finest ingredients.

Don't Forget: Vegetarian and gluten-free choices are offered every night.

188 South Main Street
White River Junction, Vermont

- - - - -

802.281.7009

- - - - -

www.elixirrestaurant.com

THE FARMHOUSE TAP & GRILL

Farm-to-table gastropub

The Farmhouse Tap & Grill is dedicated to showcasing local farms and food producers. Their menu features award-winning local burgers, comfort entrées, artisan cheeses, vegetarian options, innovative nightly specials and homemade desserts. The taproom delivers highly prized and rare beers from Vermont and farther afield. With 24 taps—always pouring selections from Hill Farmstead and Lawson's Finest Liquids—and more than 160 bottles from around the world, visiting The Farmhouse Tap & Grill is a must for craft beer fans!

Fun Fact: The Farmhouse got its name from a "name-that-restaurant" contest organized by the local alternative newspaper, *Seven Days*. The selected name fits perfectly with the restaurant's mission to bring people together in a familiar, welcoming environment to celebrate local food.

Don't Forget: Stop in on Wednesday nights for the weekly Special Happenin's, or check out their website for a lineup of events.

160 Bank Street
Burlington, Vermont

- - - - -

802.859.0888

- - - - -

www.farmhousetg.com

FOLINO'S

Wood-fired pizza adjacent to Fiddlehead Brewing

Folino's is BYOB (beer and wine only), so pick up a growler from neighboring Fiddlehead Brewing *(see page 49)* or bring in your favorite beverage. This casual restaurant serves classic as well as innovative salads and pizzas, like their famous bacon, scallop and lemon zest pizza. Folino's believes that customer participation makes for a more at-home feel and casual atmosphere, so there's limited table service. Help yourself to your own water, plates, forks, etc. They even have a freezer filled with frosted beer and wine glasses upon request. Their pizza dough is naturally leavened and hand-formed then cooked in a large wood-fired masonry oven in temperatures up to 1,000°F. So get some beer, head to Folino's and join the party!

Fun Fact: Folino's is named after Buddy Folino, owner Buddy Koerner's grandfather. Look for a picture of the restaurant's namesake in the dining room.

Don't Forget: Folino's is BYOB—that means beer and wine only. No hard alcohol is allowed!

6305 Shelburne Road
Shelburne, Vermont

- - - - -

802.881.8822

- - - - -

www.folinopizza.com

GRIFFIN'S PUBLICK HOUSE

Gastropub + beer emporium

Griffin's Publick House is one of Rutland's newest "refined but fun" casual restaurants. It features a farm-to-table seasonal menu that includes comfort dishes artfully prepared in-house. It also touts Rutland's finest microbrew tap selection, serving 20 rotating craft beers and wines from Vermont and across the country.

Fun Fact: The restaurant's motto is 'Without question, the greatest invention in the history of mankind is beer'—definitely a preview of what you'll find on a visit!

Don't Forget: Get a free kid's meal with the purchase of each adult entrée!

42 Center Street
Rutland, Vermont
- - - - -
802.772.7997
- - - - -
www.griffinspublickhouse.com

HARPOON RIVERBEND TAPS & BEER GARDEN

Fresh local beer + great food overlooking the brewery

Order up a fresh pint of Harpoon beer along with a delicious sandwich or hearty entrée in the unique brewery setting. Seating is available inside overlooking the bottling line or outside in the beer garden. If you're not sure which beer to try, order the popular sample tray, which features four 6-ounce samples of fresh Harpoon beer on tap.

Fun Fact: You'll always find something special on tap at the Riverbend Taps & Beer Garden. Beers that are available only at the brewery are constantly rotating, so you never know what interesting pilot batch is available until you come and visit!

Don't Forget: Riverbend Taps has live music every Thursday night from 6 to 9 p.m. and on the first Saturday of each month. Check out the website for details.

See page 65 for info on the brewery.

336 Ruth Carney Drive
Windsor, Vermont
- - - - -
802.674.5491
- - - - -

www.harpoonbrewery.com

LOCALFOLK SMOKEHOUSE

BBQ, burgers, Tex-Mex + craft beer

Localfolk Smokehouse is renowned for house-smoked meats cooked in their wood-fired smoker using only local northern hardwoods. They are committed to keeping their neighbors in the economic loop and so source as much as they can locally. They also partner with other businesses to organize events, such as the Vermont Music Festival, then use the proceeds to help worthy local causes. With 25 beers on tap, the bar offers a well-balanced and ever-changing assortment of drafts from the best local, regional and international breweries. Beer styles range from IPAs and dry-hopped imperial Belgians to German altbiers and weissbiers, English bitters and Belgian dubbels. Tequilas, whiskeys and bourbons are also well represented, along with other liquors and wines from around the world.

Fun Fact: The infamous Mad Mountain Tavern, a nationally recognized live music and nightlife venue, used to occupy the space where Localfolk Smokehouse is now located. The Smokehouse continues the tradition of live music today, so be sure to check out their lineup!

Don't Forget: Localfolk Smokehouse caters events with a mobile wood-fired smoker, bringing the party to you!

Waitsfield, Vermont USA

Junction of Routes 17 & 100
Waitsfield, Vermont

- - - - -

802.496.5623

- - - - -

www.localfolkvt.com

LONG TRAIL BREWPUB

Take a hike to Long Trail for a taste of Vermont

Situated on the banks of the Ottauquechee River near the historic village of Woodstock and just minutes from Killington resort, the Long Trail Visitor Center has a pub that serves a beer-friendly menu showcasing locally sourced ingredients. Inspired by the Hofbräuhaus in Munich, Germany, pub fare is served in the comfortable indoor space or seasonally on the deck. The pub always has a large selection of year-round, seasonal and brewery-only releases.

Fun Fact: Long Trail Brewing is named after the nation's oldest long-distance trail. The Long Trail is 273 miles long and more than 100 years old.

Don't Forget: Long Trail Ale has been the best-selling craft beer in the state of Vermont for more than 24 years! Visit the brewery and take a tour to see how it's made!

See page 85 for info on the brewery.

5520 Route 4
Bridgewater Corners, Vermont

- - - - -

802.672.5011

- - - - -

www.longtrail.com

During the buildout of the brewery, Allen and Jamie realized they needed a place to unwind and have a pint after work. The taproom is a warm, inviting space that welcomes all. On any given day, you'll find a broad range of people enjoying the fresh beer along with simple yet sophisticated menu items. The rotating menu of fresh, local offerings always includes vegetarian options along with house-smoked meats and light entrées at affordable prices. The taproom offers 18 draft lines, four of which are local, nonalcoholic selections. They offer growlers and kegs to go along with brewery logo ware. During the warmer months, enjoy eating and drinking in the outdoor Biergarten. The taproom is conveniently located on the Lamoille Valley Rail Trail, a four-season recreation path.

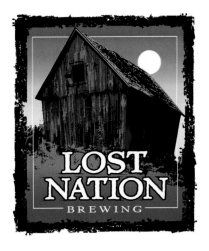

Fun Fact: The taproom used to be a railcar loading dock for the historic building where the brewery is now located back when the Lamoille Valley Railroad was active.

Don't Forget: Look for Lost Nation's limited-batch bottle releases available at the taproom.

See page 89 for info on the brewery.

See page 89 for info on the brewery.

254 Wilkins Street
Morrisville, Vermont

- - - - -

802.851.8041

- - - - -

www.lostnationbrewing.com

THE MAD TACO

Authentic Mexican fare

The Mad Taco taqueria offers some of the most authentic Mexican cuisine on the East Coast and a great craft beer selection in a casual, laid-back atmosphere. They usually have on draft two offerings from Lawson's Finest Liquids as well as two from Hill Farmstead Brewery among other local selections. And what better to go with your local craft beer than some inspired Mexican fare? Ninety percent of their food is made from scratch in their kitchens, including their own in-house smoked meats.

Fun Fact: In addition to their two locations, The Mad Taco serves food for Blackback Pub in Waterbury (*see page 161*). You'll also find their chef Eric Larson in the kitchen at the Lost Nation Taproom (*see page 181*), where he serves up an ever-changing array of creative, seasonal cuisine.

Don't Forget: The hot sauces! The Mad Taco kitchen constantly puts out fresh, new and different super-spicy sauces. And you'll never get the same one twice!

2 Village Square
Waitsfield, Vermont
802.496.3832

- - - - -

72 Main Street
Montpelier, Vermont
802.225.6038

- - - - -

www.themadtaco.com

MADISON BREWING CO.
PUB & RESTAURANT

Fermentation + civilization are inseparable

If you enjoy fine craft-brewed beer and delicious food, then visit Madison Brewing to discover what the locals already know—they are a destination for beer lovers! Choose from unique items that include pub fare and Lebanese-inspired dishes as well as homemade desserts. In the warmer months, enjoy the outdoor seating.

Fun Fact: The Madison Family owned and operated a local favorite ice cream shop called Tastee Freez for more than 40 years. Customer service, food and beverage have been integral aspects of the family's life for decades!

Don't Forget: To pick up a growler of your favorite ale while visiting the pub, and if you're into live music, be sure to visit on a Friday or Saturday night.

See page 93 for info on the brewery.

BENNINGTON, VT

428 Main Street
Bennington, Vermont

- - - - -

802.442.7397

- - - - -

www.madisonbrewingco.com

LEARN MORE ON PAKEPLATE.COM

183

THE MARSH TAVERN

Craft beer + regional cuisine in a historic location

The historic Marsh Tavern was the first building erected on the site of what is today the expansive Equinox resort. As one of the first lodging houses in Vermont, its history predates the Revolutionary War. Today it is home to a celebrated restaurant and meeting place. The Marsh Tavern is renowned for its hearty New England regional fare. The local flavors carry over into their beer selection—they offer unique brewing styles as well as seasonal offerings from Vermont and New England craft breweries.

Fun Fact: It was at the Marsh Tavern that the local Council of Safety held its first meetings and where Ira Allen, Ethan Allen's younger brother, proposed confiscating the property of the Tories to raise money to equip a regiment of the Green Mountain Boys during the American Revolution.

Don't Forget: While visiting the tavern, don't forget to try their own Equinox Ale, produced especially for the resort by Otter Creek Brewing *(see page 109).*

THE
marsh
Tavern

3567 Main Street (Route 7A)
Manchester Village, Vermont
- - - - -
800.362.4747
- - - - -
www.equinoxresort.com/dining

McNEILL'S PUB & BREWERY

Family-friendly pub with a great craft beer selection

The cozy, family-friendly pub serves homemade Mexican food (be sure to try their own chipotle-mango salsa), delicious pizzas and, seasonally, an assortment of soups. They have 14 beers on tap with plans to expand their draft capacity in the near future. Every Wednesday patrons can enjoy live Irish music.

Fun Fact: The brewery is in the basement of the pub and was once the town jail. The joke is the village drunks used to all sleep there and now they all work there.

Don't Forget: McNeill's Pub is open 365 days a year starting at 4 p.m. Monday through Thursday and at 1 p.m. Friday through Sunday. Holiday hours vary.

See page 101 for info on the brewery.

90 Elliot Street
Brattleboro, Vermont

- - - - -

802.254.2553

- - - - -

www.facebook.com/McNeillsBrewery

MULE BAR

Bringing craft beer + local food to Winooski

Located on the circle in downtown Winooski, the Mule Bar is known for serving locally sourced, thoughtfully prepared food and its amazing selection of craft beer. The bar has 16 draft lines and an extensive, diverse bottle selection that are both expertly curated and ever changing to match the seasons and the cuisine. You'll find housemade charcuterie, fine Vermont cheeses and a fantastic selection of other dishes that change according to what's available from local producers. All of this, paired with expert service in a warm and friendly atmosphere, make Mule Bar Winooski's craft beer hitching post!

Fun Fact: Besides Mule Bar's ever-changing beer selections, their bartenders are cocktail masters and there's something for everyone!

Don't Forget: Every Tuesday is Taco Tuesday— local carnitas tacos and burritos are served all day and all night!

38 Main Street
Winooski, Vermont
- - - - -
802.399.2020
- - - - -
www.facebook.com/MuleBarVT

OTTER CREEK BREWING

Pub fare + beer pairings

The pub serves light fare until 6 p.m., with menu offerings that often feature locally sourced ingredients. The taproom has a rotating selection of more than a dozen craft offerings, including some that are available only at the brewery. The brewers can often be found enjoying their shift pints and talking shop with guests, so stop in for a bite and a pint and some spirited conversation.

Fun Fact: In the summer months, a small disc golf course is set up behind the brewery for all to enjoy!

Don't Forget: You'll often find brewery-only beer releases that are only available on tap at the pub.

See page 109 for info on the brewery.

See page 109 for info on the brewery.

793 Exchange Street
Middlebury, Vermont

- - - - -

802.388.0727

- - - - -

www.ottercreekbrewing.com

LEARN MORE ON PAPBPLATE.COM

THE PARKER PIE CO.

The Parker Pie Co. started as a small pizza shop in the back of the Lake Parker Country Store *(see page 240)*. While it is still a little pizza shop in the back of a country store, the quality of the food and the community atmosphere has turned it into a favorite eatery and pub for locals and seasonal clientele alike. Their revolving beer menu features both Vermont brews and the best fermented beverages from around the globe. They are located about 20 minutes from Hill Farmstead Brewery and always have Edward APA on draft.

Fun Fact: The restaurant is housed in the building that used to be the West Glover post office.

Don't Forget: Raw oysters are shucked on premise on Fridays!

161 County Road
West Glover, Vermont

- - - - -

802.525.3366

- - - - -

www.parkerpie.com

PARKER PIE WINGS

Planes. Pies. Wings.

Parker Pie Wings brings an unexpected touch of local flair to the quiet Newport State Airport. Wings is a far cry from the terminal fast food that springs to mind when we think of airport dining. This restaurant and music venue occupies an airplane hangar that formerly served as the mechanic's workshop. Wings continues the mission of The Parker Pie Co. in West Glover by providing a place for people to gather and appreciate the Northeast Kingdom through the bounty of locally grown produce, hand-tossed pizzas and craft brews. Locals and travelers alike stop in to see what kegs are being put into rotation that day.

Fun Fact: Parker Pie Wings and Lake View Q Aviation host an annual air show in August. If world-class pizza, beer and music weren't enough, this is your chance to mix in some cool planes and stunt pilots to satisfy the whole gang. Scenic flights and lessons are also available year-round and are cheaper than you might think. Remember, "Eight hours bottle to throttle."

Don't Forget: Vermont is also home to an increasing number of distilleries producing a variety of spirits. In the tradition of supporting local breweries, Wings also features an almost complete selection of Vermont-made liquors.

2628 Airport Road
Newport, Vermont

- - - - -

802.334.9464

- - - - -

www.parkerpiewings.com

THE PERFECT WIFE RESTAURANT

Local foods, live music + diverse brews

The Perfect Wife Restaurant and The Other Woman Tavern have been serving locals and travelers for 17 years. Their commitment to locally grown and produced foods is obvious in both the menus and at the tap. In addition to carrying many Vermont-made beers, such as Foley Brothers, Long Trail, Harpoon and Switchback, they also carry a wide selection of hoppy IPAs, yeasty Belgians and rich porters from across the country. The Other Woman Tavern is the local watering hole that has become popular by offering creative food, friendly service, fresh cold beers and an eclectic wine list.

Fun Fact: Chef/Owner Amy Chamberlain was voted Vermont's Chef of the Year in 2010.

Don't Forget: The Other Woman Tavern is Manchester's place for live music with open mic most Wednesdays, live bands some Fridays and occasional acoustic performances.

The Perfect Wife
RESTAURANT

2594 Depot Street
Manchester, Vermont

- - - - -

802.362.2817

- - - - -

www.perfectwife.com

PIECASSO
Pizzeria + lounge

Piecasso Pizzeria & Lounge is a lively restaurant and bar featuring artisan, hand-tossed New York-style pizzas with a modern twist! Their newly renovated bar offers 12 rotating tap lines featuring brews from local favorites, such as Hill Farmstead, Switchback, Rock Art, Lost Nation and Trapps, as well as notable brands from across the country, like Stone, Founders, Lagunitas, Victory, Six Point and more. Piecasso is famous for its local following, friendly atmosphere and great music. They focus on local and organic ingredients and Vermont products in their creative entrées, pasta dishes, gluten-free options, kid's items, small plates, cheese and charcuterie.

- -

Fun Fact: Piecasso was one of 11 pizzerias featured in "America's Best Pizza" in *Travel & Leisure*!

- -

Don't Forget: Hit the raw bar on Tuesday nights, join in the pub-style trivia on Wednesdays and listen to local live music on Saturday nights.

1899 Mountain Road
Stowe, Vermont

- - - - -

802.253.4411

- - - - -

www.piecasso.com

LEARN MORE OR FARMPLATE.COM

191

THE PIZZA STONE VT

Baked not burnt

The Pizza Stone's Vermont-style pizza is one of a kind. This family-owned business started as a small outfit just down the road from its current location. It's now a full-service restaurant in a casual, rustic space with a full menu and 18 rotating craft beers on tap. They focus on Vermont breweries first but feature other micro-breweries from the New England region as well. The Pizza Stone VT prides itself on buying local, eating local and being local.

Fun Fact: The Pizza Stone has pefected some gluten-free selections, including pizza and their award-winning wings. They also serve gluten-free beer and local ciders to boot.

Don't Forget: They offer live local music weekly, from bluegrass to rock and roll.

291 Pleasant Street (Route 11)
Chester, Vermont

- - - - -

802.875.2121

- - - - -

www.pizzastonevt.com

POSITIVE PIE

Hand-tossed pizzas in a contemporary setting

Rooted in its beginnings as an authentic pizzeria, Positive Pie combines its old-school heritage with a new-age Vermont mindset. They are fully committed to providing a cool, contemporary, comfortable, casual urban refuge and serving hand-tossed pizzas, seasonal entrées, craft beer and cider, sustainable wines and classic cocktails.

Fun Fact: *USA Today* featured Positive Pie as one of Daymon Patterson's "10 great places for tasty takeout." Patterson is the YouTube restaurant review sensation and host of the Travel Channel show, "Best Daym Takeout."

Don't Forget: There is live music at all three locations, ranging from jazz and bluegrass to rock and hip-hop, showcasing both local talent and national acts.

22 State Street
Montpelier, Vermont
802.229.0453

- - - - -

87 South Main Street
Hardwick, Vermont
802.472.7126

- - - - -

69 Main Street
Plainfield, Vermont
802.454.0133

- - - - -

www.positivepie.com

PROHIBITION PIG

Smoked meat + libations

Prohibition Pig (fondly known by locals as Pro Pig) features one of New England's largest and best-curated selections of craft beer, proper cocktails and eclectic wines. Their extensive menu features barbecue and vegetarian and American fare. They make everything fresh in-house and use locally raised, all-natural meats, sustainably caught seafood and local produce when in season. Their knowledgeable staff will help you find the perfect drink to pair with your meal.

Fun Fact: At the time of this printing, Pro Pig had just received their federal brewer's permit. *(See page 265 for other "On Tap Soon" breweries.)*

Don't Forget: In addition to the vast array of craft beers on tap, Pro Pig has $4 Fernet drafts every day.

23 South Main Street
Waterbury, Vermont
- - - - -
802.244.4120
- - - - -
www.prohibitionpig.com

THE RED CLOVER INN & RESTAURANT

Localvore cuisine with an international flair

The farm-to-fork movement meets inspired international cuisine, and wines from around the world meet brews from around the corner at the Red Clover Inn's cozy restaurant and tavern. Chef Colin Arthur's ever-changing menu features seasonal ingredients and a fresh take on classic dishes—perfectly paired with a rotating selection of Vermont craft brews, carefully selected wines and unique cocktails. A crowd favorite is the Portuguese steak, served with duck fat fries and topped with a sizzling fried egg. The Red Clover is a beautifully restored 1840s farmhouse, just off the beaten path and only moments from Killington and Pico.

Fun Fact: The inn's sister property on Lake Champlain, the Tyler Place Family Resort, has provided all-inclusive family vacations (and plenty of Vermont beers on tap) for more than 80 years.

Don't Forget: The night doesn't have to end when dinner is over. Enjoy a drink by the fire or take a dip in the hot tub before retiring to one of the inn's welcoming rooms.

7 Woodward Road
Mendon, Vermont

- - - - -

802.775.2290

- - - - -

www.redcloverinn.com

THE RESERVOIR

Taproom with fresh eats + craft beer

The Reservoir Restaurant & Taproom is a place where the craft beer focus is on Vermont first. The goal here is to feature as many Vermont breweries as possible, including Hill Farmstead, Lawson's Finest and Waterbury's own The Alchemist. In addition to their selection of local beers, they have an extensive list of some of the finest craft brews from New England, across the country and around the world. While enjoying one of these great beers, try some of their delicious pub fare, featuring local and regional products.

Fun Fact: Mondays are Community Pint Night. The Reservoir gives each guest a wooden quarter for every beer purchased on this night. Guests can drop their quarters in the donation box, and at the end of each month, The Reservoir donates the funds to a selected charity.

Don't Forget: Enjoy The Reservoir's large outdoor seating area during the summer months.

1 South Main Street
Waterbury, Vermont

- - - - -

802.244.7827

- - - - -

www.waterburyreservoir.com

THE ROOST

Après anything bar at Topnotch Resort

Part modern lobby bar, part restaurant, The Roost is an entirely reimagined pub that serves fare with a decidedly Vermont twist. Sit outside under the pergola, by the fire pit or next to the bocce court, or lounge inside on oversized couches while sampling one of Executive Chef Steve Sicinski's small plates, perfect paired with one of The Roost's craft beers or creative cocktails.

Fun Fact: The Roost's shuffleboard table is a one-of-a-kind work of art, lovingly handcrafted right here in Vermont.

Don't Forget: The Roost rocks every night après ski with shuffleboard, shenanigans and more.

4000 Mountain Road
Stowe, Vermont

- - - - -

802.253.6471

- - - - -

www.topnotchresort.com/explore/dining/the-roost

THE SKINNY PANCAKE

On a mission to change the world

The Skinny Pancake is deeply committed to the ideals of social entrepreneurship. They believe in the importance and potential of the double bottom line: financial viability and socially responsible business practices. In keeping with this belief, they source as much as possible locally and work with more than 40 local farms and value-added producers. They are also Vermont's largest contributor to One Percent for the Planet. They feature a rotating selection of craft beer favorites from Fiddlehead, Hill Farmstead, Drop-In, Long Trail, Wolaver's and more. They also feature a full bar with many local spirits and specialty cocktails at both the Burlington waterfront and Burlington International Airport locations.

Fun Fact: Before opening the restaurant in 2006, Skinny Pancake's Church Street cart was the first vending cart ever in the Vermont Fresh Network.

Don't Forget: You can still find their vending cart seasonally stationed right on Church Street in downtown Burlington!

60 Lake Street
Burlington, Vermont
802.540.0188

- - - - -

89 Main Street
Montpelier, Vermont
802.262.2253

- - - - -

Burlington International Airport
802.497.0675

- - - - -

www.skinnypancake.com

STATION TAP ROOM

Mount Snow's slopeside craft beer bar

The après ski experience has been stepped up a few notches at Station Tap Room (located on the second floor of Mount Snow's Main Base Lodge) thanks to 18 rotating taps, including beers brewed exclusively for Mount Snow by the likes of Harpoon, Northshire and Rock Art. Be sure to check the big menu on the wall each time you visit—you'll undoubtedly see something new. Station Tap Room is open seven days a week during the winter.

Fun Fact: Fifty gallons of Harpoon Beartrap Brown, a strong Belgian brown ale brewed exclusively for Station Tap Room, was aged in bourbon barrels for a year before being served at the Mount Snow Brewers Festival.

Don't Forget: In addition to the Brewers Festival held each summer, there is a Mount Snow Winter Brewers Festival held on the first Saturday of April.

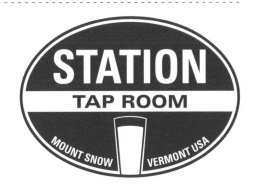

39 Mount Snow Road
West Dover, Vermont

- - - - -

800.245.7669

- - - - -

www.mountsnow.com/dining

LEARN MORE ON FARMPLATE.COM

STONE HEARTH INN & TAVERN

Rotating menu + unique ales in a converted farmhouse

The Stone Hearth Tavern has a full bar and features a rotating selection of 13 microbrew favorites. Their kitchen focuses on great yet simple foods that are complemented by savory wines and unique ales. The diverse rotating menu is inspired by local and regional ingredients.

Fun Fact: Every Sunday is Wing Night, featuring 11 varieties of wings.

Don't Forget: The tavern is family friendly, and they host an open mic every Friday.

698 Route 11 West
Chester, Vermont

- - - - -

802.875.2525

- - - - -

www.stonehearthinnvermont.com

THREE PENNY TAPROOM

Craft food for craft beer

Three Penny Taproom is home to 24 expertly curated draft lines with the finest craft beers that Vermont and the world has to offer. They harmonize the skill and care the brewers take in making these beers with artfully executed, locally sourced food. Three Penny Taproom prides itself on combining these two aspects with exceptional service. They also have a great lineup of hand-selected wines as well as creative, unusual cocktails featuring their own spirits, which they make by infusing liquors with various herbs and spices. In honor of the mastery of craft, Three Penny Taproom offers craft food for craft beer.

Fun Fact: Wednesday at TPT is now known as Weensday, as they mostly play music from, or similar to, their favorite band Ween on this day.

Don't Forget: Draft lines change literally by the minute. It's best to either be a local or live like one so as not to miss any special selections.

108 Main Street
Montpelier, Vermont

- - - - -

802.223.8277

- - - - -

www.threepennytaproom.com

TROUT RIVER BREWING COMPANY PUB

Trout River Brewing Company is a pizza place and brewery all in one. Step up to the counter and order one of their delicious hand-tossed pizzas along with a signature housemade brew. The pizza dough is made with a 17-year-old sourdough culture.

Fun Fact: You'll find Trout River brewmaster Dan Gates behind the bar most Friday and Saturday nights talking to customers and pouring beers.

Don't Forget: Before you leave, remember to grab a growler to go filled with one of their 10 delicious beers on tap. (Only three of their beers can regularly be found in bottles.)

See page 131 for info on the brewery.

Brewed in Vermont • Naturally Unfiltered

645 Broad Street
Lyndonville, Vermont

- - - - -

802.626.9396

- - - - -

www.troutriverbrewing.com

An American gastropub

Twiggs is a unique pub, restaurant and entertainment venue located in the revitalized downtown of St. Albans. It's a comfortable place to catch a sporting event, unwind with business colleagues or just relax with friends and family and enjoy one of their ten locally brewed craft beers on tap.

Fun Fact: The beer cheese sauce is made with local St. Albans brewery 14th Star's beer.

Don't Forget: Twiggs serves a late-night menu on Friday and Saturday and has live entertainment every weekend. Support local talent at their open mic night on the first Wednesday of the month.

24 North Main Street
Saint Albans, Vermont
- - - - -
802.524.1405
- - - - -
www.chowbella.us

203

VERMONT ALE HOUSE

Always Ale House responsibly

Vermont
Ale House

The Vermont Ale House offers locals a fun and friendly atmosphere to enjoy delicious food, rare beer and inventive cocktails. Inspired by The Shed, Stowe's legendary bar that closed in 2011, the 40-foot-long copper bar is the heart of the restaurant, allowing patrons to belly up, hang out and mingle with friends old and new. Cocktails are inspired by the Prohibition era and are prepared by the award-winning mixologist and bar manager Kevin Baker. The beer served at the Vermont Ale House changes daily. They receive unique selections based on availability from brewers around the world, so make sure to try it if they have it!

Fun Fact: Some believe the Vermont Ale House is haunted. The restaurant is home to a library of rare books and magazines dating back more than 100 years. Sometimes books are mysteriously moved from one location to another overnight!

Don't Forget: Be sure to order the Rumors & Lies cocktail. The drink was designed by in-house mixologist Kevin Baker and is inspired by the small-town culture where news spreads fast. He uses pyrotechnics to create this delicious concoction.

294 Mountain Road
Stowe, Vermont

- - - - -

802.253.6253

- - - - -

www.vermontalehouse.com

THE VERMONT PUB & BREWERY

Pioneers of Vermont craft beer

The Vermont Pub & Brewery, Vermont's original brewpub, opened in the heart of downtown Burlington in 1988. The original vision of offering high-quality, handcrafted beers and reasonably priced, wholesome foods in a family-friendly atmosphere remains at the forefront of the pub's mission today. VPB is firmly embedded in the community and partners with more than 20 local food producers to offer artisan cheeses, fresh Angus beef and homemade cheesecake amongst other local items. The kitchen also sometimes offers homemade breads made from spent grains from the brewery cellars.

Fun Fact: VPB has served more than 150,000 fish and chips dinners since its inception and over 700,000 burgers made from locally raised Angus beef from the LaPlatte River Farm.

Don't Forget: Each year, over 60 styles of freshly brewed beer are offered at the pub! The 15 taps offer regularly brewed ales and lagers along with an ever-changing selection of seasonal varieties. Veteran brewmaster Russ FitzPatrick produces a new beer every Monday in the small experimental brewery.

See page 135 for info on the brewery.

144 College Street
Burlington, Vermont

- - - - -

802.865.0500

- - - - -

www.vermontbrewery.com

WHETSTONE STATION RESTAURANT

Waterfront dining + rooftop Biergarten

Whetstone Station Restaurant and Brewery combines a very small three-barrel brewery with a large casual restaurant and open-air rooftop Biergarten, located right on the Connecticut River in a beautiful waterfront location. They offer inspired pub fare, with everything from small bites to full entrées. Along with one always changing unique craft beer of their own, Whestone Station offers dozens of guest craft beers on tap as well as specialty drinks. The unique round building has been restored and is now a beautiful firelit space with a contemporary, industrial vibe.

Fun Fact: The building was originally a gashouse, hence its unique round shape. It stored and provided fuel for Brattleboro's street lamps as early as the end of the 18th century.

Don't Forget: Whetstone Station has limited parking in front, but there's more across the street by the train station, and the parking garage is connected via the River Walk pathway.

See page 139 for info on the brewery.

36 Bridge Street
Brattleboro, Vermont

- - - - -

802.490.2354

- - - - -

www.whetstonestation.com

WINDSOR STATION RESTAURANT

American neighborhood dining + craft beer

Tour some of the best Vermont breweries by sampling beers from Windsor Station's 10 taps. Sit in the bucket seats at the bar made from local wood, relax in the cozy booths or dine by candlelight in one of the dining rooms. The restaurant not only supports the Vermont beer scene, but also the state's craft distilleries and farms. You'll find a creative cocktail list featuring local vodka, whiskey and gin. Vermont beef burgers, homemade pastas and local produce and cheeses are also abundant throughout the menu.

Fun Fact: This historic train station welcomes the Amtrak twice a day, so time your visit accordingly if you want to hear the whistle and feel the rumble! The conductor blows his whistle twice as many times when he sees a full station!

Don't Forget: It's a good idea to call ahead for reservations in the dining room as it fills up quickly. The barroom is walk-in only.

26 Depot Avenue
Windsor, Vermont

- - - - -

802.674.4180

- - - - -

www.windsorstationvt.com

WORTHY BURGER

Craft beer + burger bar

This top craft beer destination features an ever-changing list of 15 drafts with a focus on Vermont's most sought-after artisanal breweries and some of the best small breweries in New England. All styles are covered, with a nice blend of session beers and imperials. Executive Chef Jason Merrill is always searching for the "worthy difference" in everything he prepares—something elemental that turns his high-quality locally raised ingredients into a special experience. The Worthy Burger has a simple, creative menu, including an assortment of burgers and hand-cut fries that are cooked twice in beef tallow. They serve high-quality local food and drink at affordable prices by dealing directly with farmers and offering a simple menu that produces almost no waste.

Fun Fact: Join the Worthy Burger 50/50 Club: Drink 50 beers and eat 50 burgers (you have a year to accomplish this feat) and receive a Worthy hat and lifetime membership in the Worthy VIP Club.

Don't Forget: The Worthy Burger only accepts cash, though an ATM is available on the premises.

56 Rainbow Street
South Royalton, Vermont

- - - - -

802.763.2575

- - - - -

www.worthyburger.com

WORTHY KITCHEN

Craft beer + test kitchen

Worthy Kitchen has an eclectic menu and an 18-beer draft list focused on the best small breweries of Vermont and New England, along with a few gems from away. The restaurant also has seven wines on draft as well as hard cider and Rookies Root Beer to keep designated drivers from complaining too loudly. All liquid offerings are served in the Worthy Glass, specially designed to preserve volatile flavor and aromas. The Worthy staff knows beer and are more than ready to help pair your drink with something from the menu.

Fun Fact: Worthy Kitchen is the commissary for Worthy Burger and the headquarters of Worthy Catering, acting as a test kitchen for refining current recipes, creating new dishes and exploring future projects. It's also the heart of the Three Dudes's ever-evolving relationship with farmers as their operation continues to expand into the fields around them and blurs the line between farm and table.

Don't Forget: On Taco Tuesdays, the Food Dude explores his southwestern roots while the Beer Dude taps a special keg at 5 p.m. Worthy Kitchen also serves brunch on Saturdays and Sundays. Be sure to try the Worthy Bloody Mary!

442 Woodstock Road
Woodstock, Vermont

- - - - -

802.457.7281

- - - - -

www.facebook.com/WorthyKitchen

BUY

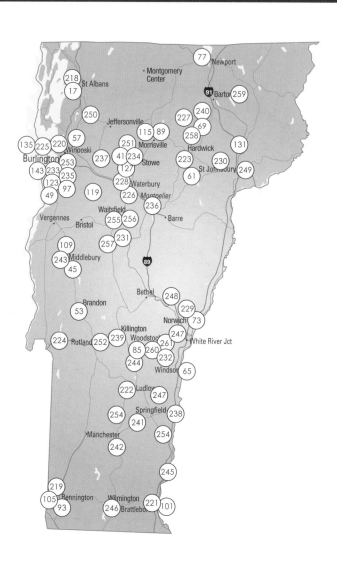

BUY

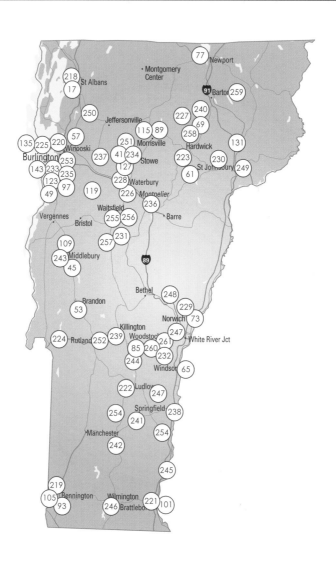

77 Newport

• Montgomery
  Center

218 St Albans
17

91 Barton 259

250   Jefferssonville   227 240
                        69
57   115 89   258
135 225 220 251 Morrisville   Hardwick   131
Burlington 253 237 41 234 Stowe   223   230
143 233 235   127   61 St Johnsbury 249
123 97
49   119   228 Waterbury
               226 Montpelier
            236
        Waitsfield
        255 256   Barre
        257 231

109
243 Middlebury   89
45

        Bethel   248
Brandon   229
53   Norwich 73
        Killington 247
224 Rutland 252 239 Woodstock 261 White River Jct
               85 260 232
               244
                    Windsor 65

            222 Ludlow
                247
            254 Springfield
                241 238
        Manchester 254
            242

                245

219
105 Bennington Wilmington
93   246 Brattleboro 221 101

BUY

AS THE CROW FLIES

Your local kitchen store + craft beer destination

Established in 1998, As the Crow Flies is the ultimate kitchen store. With a strong emphasis on locally made goods, much thought has been put into the well-stocked store's selections. There is something to be found for all those culinarily inclined—cooks, bakers, wine and beer enthusiasts and foodies. In addition to the array of wines from around the world, the store has an interesting selection of craft beers focused heavily on local Vermont brews. Their Vermont beer and cheese gift baskets are a popular gift item and a great way to sample the local flavors.

Fun Fact: As the Crow Flies was the first store in Vermont to stock the hard-to-find beers from 14th Star Brewing Co. *(see page 17)*, based right in St. Albans, Vermont!

Don't Forget: Stop in every fourth Friday of the month from 4 to 7 p.m. for a free tasting of an ever-changing lineup of beers and wines.

58 North Main Street
St. Albans, Vermont

- - - - -

802.524.2800

- - - - -

www.asthecrowfliesvt.com

THE BEVERAGE DEN & SMOKESHOP

One of the largest craft beer selections in the region

The Beverage Den has been a beer connoisseur's paradise for nearly 40 years. The Den offers well over 100 different craft beers, featuring many local Vermont and New Hampshire brewers as well as national producers spanning from California to Maine. The Den also stocks a growing number of gluten-free beers. Specialty kegs are available, as well as a vast array of homebrewing supplies, wines, hard ciders, snacks and gift items.

Fun Fact: The beer manager Shane, who has cheerfully sampled every beer in the store, would be happy to share his experiences with you!

Don't Forget: They also have a large selection of fine wines and Vermont gift items.

THE BEVERAGE DEN *and* **Smokeshop**

340 North Street
Bennington, Vermont

- - - - -

802.442.2861

BEVERAGE WAREHOUSE

Step outside the beer box

You know Beverage Warehouse is not your mundane liquor store when you notice the store's managers have titles such as Tactical Hop-erator, Libation Hooligan, Company Ogre and Rockstar. Their addiction to the art of hunting down gem beers is obvious the moment you walk into this old landmark warehouse. First you see the copper-topped growler bar, then you walk into the beer oasis with more than 1,000 beers awaiting your selection. The knowledgeable staff will help you explore your palate and step outside your beer box.

Fun Fact: The building initially housed one of the first screen manufacturing plants in the United States. It later became a concrete block manufacturing facility prior to its transition to a beverage store in the late 1970s. The founder literally blasted walls with explosives to create some of the doorways that now exist, and the walls are still jagged in places! In the three decades the Swiatek family has owned the Bevie, there has been as much of the original structure and feel maintained as possible, including baseball scores from the 1970s and 1980s World Series written all over the walls.

Don't Forget: The Bevie has a vintage beer cave where beers that benefit from some aging are held and released. The store also hosts frequent free beer, wine and liquor tastings!

1 East Street
Winooski, Vermont
- - - - -
802.655.2620
- - - - -
www.beveragewarehousevt.com

BRATTLEBORO FOOD CO-OP

Member-owned, member-governed

The Brattleboro Food Co-op was established in 1975 as a small buying club. It has grown into a bustling natural foods market and deli that carries more than 3,240 different local products! The co-op is housed downtown in a beautiful 14,580-square-foot building that has many environmentally friendly features, including solar panels on the roof and a system of using waste heat from its refrigerators and freezers to heat the store as well as neighboring apartments. The co-op is community-owned by more than 6,000 active shareholders, and their goal is to be as accessible and responsive to customers as possible. They strive to provide quality foods at reasonable prices in a relaxed and friendly environment.

Fun Fact: Through their extensive education and outreach efforts in local schools and community centers, the co-op staff works with more than 1,900 children a year, exploring nutrition, local foods and cooking fun.

Don't Forget: The co-op carries more than 100 types of craft beer, including selections from many local brewers. Tastings are held throughout the year at the store, and there is a popular annual holiday wine and beer tasting with scrumptious food pairings the second Friday in December.

2 Main Street
Brattleboro, Vermont

- - - - -

802.257.0236

- - - - -

www.brattleborofoodcoop.com

BREWFEST BEVERAGE

Take good beer home

The enthusiasm and excitement about craft beer at Brewfest Beverage is what drives this store to offer an outstanding selection of craft beer from Vermont, the United States and all over the world. The staff believes in local production and craftsmanship, and they work hard to bring in the most interesting, highest quality and hard-to-find limited releases available. Vermont has many outstanding microbreweries as well as topnotch cider and wine producers. Brewfest Beverage is proud to promote these craft producers with a high-profile showcasing of the quality, local products available.

- -

Fun Fact: Brewfest Beverage is family owned and operated. The great music that enhances the atmosphere at the store comes from the family's personal collections!

- -

Don't Forget: Brewfest Beverage pours growlers all day, every day, and they have beer tastings on Saturdays throughout the ski season!

199 Main Street
Ludlow, Vermont

- - - - -

802.228.4261

- - - - -

www.brewfestbeverage.com

BUFFALO MOUNTAIN
FOOD CO-OP

Food for people not for profit

The Buffalo Mountain Food Co-op is a not-for-profit corporation owned by its members. Their purpose is to provide their membership and the greater community with whole foods and other products that are grown or made with the health and well-being of the planet and its inhabitants in mind. They are committed to making these items available at as low a cost as possible by involving their members directly in the daily workings of the store and by offering "food for people not for profit." They carry a small but ever-changing selection of Vermont and regional beers and wines, and they are very open to suggestions from customers as to which new producers to try.

Fun Fact: Buffalo Mountain is one of the oldest co-ops in the state, and the only one to still use a collective form of management.

Don't Forget: Bring your own bag(s) for 5¢ off your bill per bag. Look for periodic wine tastings (and sometimes beer tastings), especially around the holidays and during First Friday celebrations in the summer.

Buffalo Mountain
Food Co-op & Café

39 South Main Street
Hardwick, Vermont

- - - - -

802.472.6020

- - - - -

www.buffalomountaincoop.org

CASTLETON VILLAGE STORE

Big city beer selection in a small country store setting

The store's location in a college town means they have to be on their toes with the constantly evolving microbrew environment—new trends and changing tastes keep their selection fresh, up to date and ever expanding. They even bring in seasonal and one-time brews that are available only for a limited time. The store not only represents local craft brewers, but they also have a huge selection of craft beers from around the country.

Fun Fact: In addition to craft beer, Castleton Village Store also has a great selection of Vermont wine, cider and mead.

Don't Forget: Special orders are welcome.

583 Main Street
Castleton, Vermont

- - - - -

802.468.2213

- - - - -

www.castletonvillagestore.com

CITY MARKET

Onion River Co-op

City Market Onion River Co-op is a consumer cooperative with more than 9,500 members. They sell wholesome foods and other products while working to build a vibrant, empowered community and a healthier world, all in a sustainable manner. Located in downtown Burlington, City Market provides a large selection of organic and conventional foods, including thousands of local and Vermont-made products. They have a diverse selection of Vermont beers and wines, and their knowledgeable, friendly staff will help you make the perfect choice!

Fun Fact: The market is open to everyone, but there are many benefits of membership, including earning money back from purchases through the annual patronage refund program.

Don't Forget: Sign up online for their beer and wine department e-newsletter.

City
Market
Onion River Co-op

82 South Winooski Avenue
Burlington, Vermont

- - - - -

802.861.9700

- - - - -

www.citymarket.coop

CRAFT BEER CELLAR
OF WATERBURY

Waterbury's one-stop craft beer shop

CBC's goal is to offer customers the most sought-after craft beers and to cultivate the craft beer community by offering tastings, hosting educational programs in their classroom space and encouraging customer requests and feedback. Striving to bring you the latest releases from around the world with a focus on local breweries, CBC Waterbury sells bottles and cans, six packs and bombers, plus a full selection of homebrewing supplies and equipment. You can expect topnotch customer service from their knowledgeable staff.

Fun Fact: CBC Waterbury offers a counter-pressure growler system, which helps duplicate the environment inside of a keg in your growler, keeping your beer fresher longer.

Don't Forget: Follow them on Twitter @cbc_waterbury for information on tastings and new releases. And remember to always bring your growler back clean.

3 Elm Street
Waterbury, Vermont
- - - - -
802.882.8034
- - - - -
www.craftbeercellar.com

CRAFTSBURY GENERAL STORE

Good food beckons good beer

This small market in the village of Craftsbury has all the quintessential Vermont appeal one would expect of a New England general store, with a local Craftsbury twist. Locals and visitors alike come by for fresh greens grown right down the road, gently pasteurized milk from Sweet Rowen Farmstead and tasty sandwiches and pizzas from their deli. And, what goes better with local food than great local beer. The store is nestled in a location where some of the best beer is made just miles away. Their customers are always looking for a new craft brew to try, so they keep an extensive rotating beer selection and also have a growler filling station.

Fun Fact: You'll be hard-pressed to find a sillier staff around . . . and a Shih Tzu unofficially owns the store!

Don't Forget: Globe Trotting Wednesdays are the perfect night to pick up a take-out dinner featuring food from around the world made with Vermont ingredients.

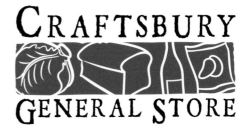

118 South Craftsbury Road
Craftsbury, Vermont

- - - - -

802.586.2440

- - - - -

www.craftsburygeneralstore.com

CROSSROADS BEVERAGE & DELI

The one-stop shop

Crossroads Beverage & Deli is known for having a great selection of quality products. Their deli has fresh sandwiches, pizzas and homemade soups made daily (they usually have nine!), and their beer selection is one of the largest in the state! They have 14 ice-cold coolers of beer—six of which are devoted to craft beers—and a selection of 50 cold bombers, with 20 more feet of bombers on the shelves! Crossroads also has a growler bar with six craft beers on tap, and they have a variety of kegs in stock. Crossroads has spent many years building relationships with breweries and sales reps to make sure customers get the most limited, hard-to-find beers out there. The same dedication that goes into their beer selection also goes into their diverse wine selection. Crossroads is also a Vermont liquor outlet where you can find many products from the state's innovative distilleries.

Fun Fact: Crossroads Beverage has been locally and family-owned for almost four decades.

Don't Forget: Check their Facebook page for new beer listings, wine discounts and deli specials.

52 North Main Street
Waterbury, Vermont

- - - - -

802.244.5062

DAN & WHIT'S

"If we don't have it, you don't need it."

Dan & Whit's is a true Vermont general store that has been run by the same family for three generations. Their goal is to provide daily necessities to customers in the Upper Valley as well as stocking their shelves according to customer requests. They pride themselves on personal service and caring, and so you can usually find what you're looking for! They are firmly committed to supporting the local economy and stock many local brews from Vermont, New England and beyond.

Fun Fact: One couple was married here!

Don't Forget: Dan & Whit's is open every day from 7 a.m. to 9 p.m., except for Thanksgiving and Christmas, when the store closes at noon.

319 Main Street
Norwich, Vermont

- - - - -

802.649.1602

- - - - -

www.danandwhits.com

DIAMOND HILL STORE

On the Green in Danville

Diamond Hill Store is a specialty food shop located in Vermont's picturesque Northeast Kingdom. Over the years, the store has become known to locals and visitors alike as one of the hidden treasures of the Northeast Kingdom. So much more than a boutique wine shop, Diamond Hill is a place where charming small-town character comes together with a world-class selection of wines, beers, cheeses, old-fashioned candies and other specialty foods. Their beer selection reflects the whole new genre of educated and demanding beer drinkers who are passionate about artisanal beers.

Fun Fact: The store prides itself on its cheese case, which features 30 wheels of both Vermont and European cheeses that the staff cuts personally for customers. Free tastings are offered!

Don't Forget: Diamond Hill Store is the only state-licensed liquor store in the area.

11 Route 2 East
Danville, Vermont

- - - - -

802.684.9797

- - - - -

www.diamondhillstore.com

EAST WARREN
COMMUNITY MARKET

Think local, shop local, eat local

It may be small, but the market sure packs a lot in! From the finest locally sourced vegetables, breads, eggs and meats to a diverse selection of beer and wine, customers are sure to find all they need. The market has an ever-changing and growing beer assortment, with more than 70 varieties at any given time from breweries all over the country, including the tasty and elusive Heady Topper, produced by The Alchemist *(see page 21)* just 20 miles away. The market is fortunate to have had a relationship with The Alchemist since the early days, and it is known for having Heady Topper available long after other places have sold out.

Fun Fact: As a co-op, the market is 100 percent customer supported. After opening just four years ago, it wasn't always an easy road. But today, with the tremendous support of the community, the market is fully stocked and open seven days a week.

Don't Forget: Every Friday night the market prepares a family-style dinner for pickup that feeds four to six and pairs well with the selection of craft brews.

EAST WARREN
COMMUNITY
MARKET

42 Roxbury Gap Road
Warren, Vermont

- - - - -

802.496.6758

- - - - -

www.eastwarrenmarket.com

F.H. GILLINGHAM & SONS GENERAL STORE

Your money's worth or your money back

This traditional Vermont general store is located in the picturesque and historic village of Woodstock. They offer a first-rate selection of Vermont maple syrup and other maple products, Vermont cheddar, local, domestic and imported cheeses and gourmet and Vermont specialty foods in addition to a complete line of housewares, hardware, sporting and fishing supplies, gardening tools, classic children's toys and the finest wines and beers available.

Fun Fact: Gillingham's is one of the first general stores in Vermont to offer more than 200 individual selections of beer from Vermont and around the world!

Don't Forget: Don't forget to visit with the resident cats, Ada and Sara—the store's official heads of public relations!

16 Elm Street
Woodstock, Vermont

- - - - -

802.457.2100

- - - - -

www.gillinghams.com

GUILD FINE MEATS

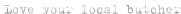

Love your local butcher

Located right on City Hall Park in the heart of Burlington, Guild Fine Meats is a neighborhood delicatessen and butcher shop. Open for breakfast and lunch, the market serves sandwiches featuring housemade deli meats and breads, party platters, soups, salads and baked treats. You can also pick up dinner—choose from a variety of housemade Vermont meats, such as sausages, smoked ham and turkey, roast beef, marinated steaks, whole trussed chickens, seasoned chops, burgers and more. Jeff Baker, manager of the bar program at The Farmhouse Tap & Grill, stocks the shelves with the best craft beers from Vermont and around the globe. His interesting selections are food-friendly and pair well with Guild's meat and cheese offerings.

Fun Fact: Guild makes their own deli meats from Vermont beef, pork and turkey.

Don't Forget: Stop in early for one of their breakfast sandwiches or a maple-glazed donut!

111 St. Paul Street
Burlington, Vermont

- - - - -

802.497.1645

- - - - -

www.guildfinemeats.com

Harvest Market is a one-stop gourmet shop owned by Donna Carpenter, president of Burton Snowboards. Donna once worked for Ina Garten of Barefoot Contessa fame, and this inspired her to open what is now a Stowe favorite for locals and visitors alike. Stop in for some of the best coffee in town and delicious specialty foods to go—choose from a large selection of salads, sandwiches, entrées, baked goods and wood-fired artisan breads, all prepared by their chefs and bakers. The market also offers a wonderful selection of cheeses, meats, farm-fresh produce, wholesome staples and an interesting, curated selection of wine and beer.

Fun Fact: Hundreds of pounds of masonry sand are stored on top of the ovens to help them stay hot during the cold Vermont winters!

Don't Forget: Harvest Market is committed to selling as many local products as they possibly can!

harvest
market

1031 Mountain Road
Stowe, Vermont
- - - - -
802.253.3800
- - - - -
www.harvestatstowe.com

HEALTHY LIVING MARKET

Taking care of people

- -

Healthy Living Market and Café is a wonderful source for local, organic, all-natural and just plain delicious food and drinks from the best, most-trusted local, national and international sources. Their beer department is stocked with Vermont favorites like Switchback and Heady Topper, along with limited-edition bottle releases, special brewer collaborations and seasonal specialties. They are also happy to try to fulfill special orders—just let them know and they'll try to get it for you!

- -

Fun Fact: Healthy Living also carries a large selection of gluten-free beer, such as Celia Saison and international favorites like Estrella Damm Daura.

- -

Don't Forget: Their extensive cheese department is the perfect place to find pairings for your favorite brews. Whether serving stout or summer ale, their cheese experts will help you find a selection that perfectly complements your beer.

Market and Café

222 Dorset Street
South Burlington, Vermont

- - - - -

802.863.2569

- - - - -

www.healthylivingmarket.com

HUNGER MOUNTAIN COOP

Your community-owned natural market + café

Hunger Mountain Coop is nestled on the banks of the Winooski River in Montpelier, the state capital. This member-owned cooperative offers a wide selection of natural foods and products, as well as a full-service café with prepared foods, including a hot bar, salad bar and sandwich and burrito menu. The coop is more than just a place to buy healthy food—it is a community resource that promotes nutritional awareness, local sustainability and environmental responsibility. The beer aisle is home to dozens of offerings that will delight both those new to the exciting world of beer as well as the most sophisticated beer drinker. You'll find the best Vermont microbreweries well represented, as well as specialty, seasonal, regional and international offerings.

Fun Fact: The coop celebrated its 40th anniversary in 2013 and marked the occasion with a collaboration beer brewed with Lawson's Finest Liquids.

Don't Forget: Anyone can shop at the coop, member-owner or not.

623 Stone Cutters Way
Montpelier, Vermont

- - - - -

802.223.8000

- - - - -

www.hungermountain.coop

JERICHO CENTER COUNTRY STORE

In continuous operation for more than 200 years

Since 1807, this historic Vermont country store has served as the hub of Jericho Center, a small town made famous by being home to Wilson "Snowflake" Bentley. It is a true Vermont country store with groceries, a gourmet deli, a working post office, Vermont gifts, a full line of beer and wine and much more! Walk through the door of this working State Historic Site and step back in time. As you wander the aisles, relics of the past are on display, including an antique hoosier, a checkerboard, pickle barrels, vintage post office boxes and century-old signs and advertisements. Its locally famous deli and grill features Boar's Head meats and cheeses, paninis, LaPlatte River Farm beef burgers and hand-cut fries among other delicious offerings.

Fun Fact: According to the Vermont Grocers' Association, Jericho Center Country Store is Vermont's longest running country store—operating continuously since 1807!

Don't Forget: The country store offers a full array of Vermont brews, including The Alchemist's Heady Topper, Switchback, Citizen Cider, 14th Star Brewery, Rock Art, Trout River, Foley Brothers and dozens more national craft beers.

25 Jericho Center Circle
Jericho Center, Vermont

- - - - -

802.899.3313

- - - - -

www.jerichocountrystore.com

JOE'S DISCOUNT BEVERAGE

Wine, beer + spirits

Joe's Discount Beverage is all about variety. The staff works hard to get new and rare beers as soon as they're available—whether it's a new IPA offering or a tried-and-true stout. The store stocks beer from the big craft breweries as well as some rarities you might not have seen on other shelves. It also offers a constantly rotating selection of growlers so customers have access even to beers that aren't necessarily available in bottles. Joe's values selection and quality above all else for the beer kept in stock.

Fun Fact: The owner's family created Woodchuck Cider!

Don't Forget: Joe's also carries an ever-expanding inventory of homebrewing supplies for both extract and all-grain brewers.

335 River Street
Springfield, Vermont

- - - - -

802.885.3555

KILLINGTON MARKET

Killington's on-mountain grocery store

Killington Market is the area's on-mountain grocery store offering an extensive selection of beer, cider, wine and mead. The microbrew selection is constantly expanding, and it includes a full line of Vermont and New England beers such as family-owned and operated Rock Art Brewery and Maine Beer Company, an innovative brewery that donates all grain and yeast by-products to local farmers. The market also has hard-to-find West Coast beers because of a unique relationship with one of its distributors. This full-service grocery store and delicatessen offers breakfast, lunch and dinner to go.

Fun Fact: The store has been continuously open for almost four decades!

Don't Forget: You'll find a wide variety of Vermont specialty foods, including cheeses, meats and maple products—perfect for pairing with your craft brews.

killington market

2023 Killington Road
Killington, Vermont

- - - - -

802.422.7736

- - - - -

www.killingtonmarket.com

LEARN MORE ON PAMPLATE.COM

239

LAKE PARKER
COUNTRY STORE

Tell 'em Merle sent you

This small country store shares a building with Parker Pie Co. *(see page 188)* in the village of West Glover. Collectively they are a gathering place for their community, and they strive to share this feeling with visitors as well. They firmly believe in supporting their community and a sustainable Vermont—the goods you will find are local and organic whenever possible. They are surrounded by farmers who care, and because of this, they are able to offer fairly priced local meats, eggs, dairy and produce throughout the year. The grocery items they carry are a mixture of convenience and specialty items with a health-minded theme. Many of the staff are beer geeks, and so not surprisingly, their beer selection is constantly changing and is as eclectic as the small space can accommodate. They also have a solid but affordable wine selection.

Fun Fact: Lake Parker County Store and the post office that preceded it have occupied the current building since before the Civil War.

Don't Forget: Lake Parker Country Store also serves as a venue for local arts and crafts, including beautiful local jewelry and pottery.

161 County Road
West Glover, Vermont

- - - - -

802.525.6985

- - - - -

www.facebook.com/vermontcountrystore

MEDITRINA

Health, longevity + wine

Meditrina's store selection is ever-changing as they continue to taste and select new wines, beers and cheeses to share with customers. They handpick the inventory for a variety of palates and for every price range. You will always find something new on the ever-expanding beer wall including special seasonal brews and an assortment of craft beers. They have a great selection of beer made locally as well as many favorites from popular breweries throughout the United States and many of the classics from Germany, Belgium, France and beyond.

Fun Fact: Meditrina is named after the Roman goddess of health, longevity and wine!

Don't Forget: A fine selection of cured meats, speciality crackers, chutneys, preserves, chocolate and more is also available.

177 South Main Street
Chester, Vermont

- - - - -

802.875.9463

- - - - -

www.meditrinavt.com

LEARN MORE ON FARMPLATE.COM

MEULEMANS' CRAFT DRAUGHTS

A mom + pop beer shop

Meulemans' is an intimate shop specializing in Vermont and regional beers. To round out the offerings, they have an extensive selection of European and American beers from farther afield. They fill growlers from their two taps featuring local beers, many of which are only available on draught. They also sell hard cider, mead, wine and artisanal cheeses along with many other Vermont products.

Fun Fact: Right next door is Tony's Pizza, which serves authentic New York-style pizza. They are BYOB, so you can take great beer with you to pair with their delicious pizza.

Don't Forget: Wednesday is Growler Sale Day so be sure to have your growler with you whether skiing, riding, hiking or biking nearby.

Junction of Routes 100 & 30
Rawsonville, Vermont

- - - - -

802.297.9333

- - - - -

www.craftdraughts.com

MIDDLEBURY NATURAL FOODS CO-OP

Your locally grown, community owned co-op since 1976

Middlebury Natural Foods Co-op is member-owned and dedicated to providing wholesome natural foods and environmentally sound household products at the best prices possible. They strive to set prices that are affordable for the consumer while fair to the growers and producers. They offer personal service and a friendly community atmosphere for all shoppers. Beer isn't so much a department at MNFC as a passion. The staff has wide-ranging tastes, loves to talk beer and will always know the right staff person to ask about specific beer styles. The co-op's overall philosophy is to support local and small producers, and this philosophy is evident in the beer selection. They also carry a wide range of award-winning national and international beer. MNFC has one of the best selections of beer in Addison County.

Fun Fact: MNFC sells beer from Stift Engelszell, one of only 10 Trappist breweries in the world, and the only Trappist monastery in Austria.

Don't Forget: Membership at the co-op has lots of advantages, but remember, "Member or not, anyone can shop at the co-op."

9 Washington Street
Middlebury, Vermont
- - - - -
802.388.7276
- - - - -
www.middleburycoop.com

PLYMOUTH ARTISAN CHEESE

The retail store at Plymouth Artisan Cheese features Vermont specialty foods along with a great selection of Vermont-crafted beers to complement their handmade cheeses. The raw whole cow's milk cheeses they produce are still made by hand in small batches following the granular curd recipe developed more than 120 years ago by the original Plymouth Cheese Factory. This granular cheese is noted for its smooth, creamy texture and rich aftertaste. All of the milk comes from a dedicated Holstein herd located in nearby Tinmouth. Plymouth Artisan Cheese is a short, scenic drive from Long Trail Brewing *(see page 85)*.

Fun Fact: Plymouth Artisan Cheese was started in 1890 as The Plymouth Cheese Factory by former President Calvin Coolidge's father, John Coolidge. It is the second oldest cheesemaking operation in the United States. Plymouth Cheese disappeared from the American landscape until 1962 when John Coolidge, the sole surviving child of President Calvin Coolidge, decided to return to Vermont and revive the Plymouth Cheese Factory.

Don't Forget: The retail store hosts beer tastings throughout the summer months. Check Facebook for event schedules and updates.

106 Messer Hill Road
Plymouth Notch, Vermont

- - - - -

802.672.3650

- - - - -

www.plymouthartisancheese.com

PUTNEY FOOD CO-OP

Community owned, natural + delicious

Putney Food Co-op is an independent grocery and deli that focuses on fresh, local flavors. It has served its members and the community at large for more than 72 years by providing a great selection of local meats, cheeses, snacks, beers and wines in a quaint, rural store. The craft beer selection reflects the co-op's commitment to supporting independent producers, and most of their offerings rotate on a seasonal basis. They even offer a respectable selection of gluten-free beers.

Fun Fact: Last year local purchases accounted for 37 percent of the co-op's total cost of goods, demonstrating their commitment to the local economy.

Don't Forget: The co-op is open seven days a week until 8 p.m. (9 p.m. in the summer!) for all of your beer-buying needs.

8 Carol Brown Way
Putney, Vermont

- - - - -

802.387.5866

- - - - -

www.putneycoop.com

RATU'S
Liquor + market

Ratu means "king" in the Fijian language, and if you're in the market for Vermont brews, you'll surely be treated like royalty when entering Ratu's Liquor & Market. Much of their focus is on supporting local. They carry all of the Vermont-made liquors, local ciders and an excellent variety of Vermont craft beers. They also have a growing selection of local wines and cigars. The most recent addition to the market is a set of growler taps featuring rotating local craft beers. Stop in and chat while having your growler filled with some of the finest local brews!

Fun Fact: Ratu's has a relaxed island-vibe atmosphere. An eclectic music selection is always playing, and customers are often caught dancing around the store!

Don't Forget: Ratu's occasionally has fun-filled beer, wine and liquor tastings to share new products.

34 West Main Street
Wilmington, Vermont

- - - - -

802.464.2252

- - - - -

www.facebook.com/ratusliquor

SINGLETON'S

General store famous for its smoked meats

Singleton's first opened in 1946 and has been smoking meats ever since. Both stores have a smokehouse on-premises where they smoke bacon, ham, cheese and salmon. Singleton's also offers a full line of conventional and specialty grocery items as well as footwear, clothing and sporting goods. Singleton's General Store in Proctorsville also features a beer and liquor cave.

Fun Fact: Singleton's won the Vermont Farmstead 2013 Mac & Cheese Challenge featuring their very own smoked ham, and the market has barely been able to keep up with demand for this cheesy comfort food.

Don't Forget: You can also buy their house-smoked meats and other local Vermont specialty foods on their website.

356 Main Street
Proctorsville, Vermont
802.226.7666

- - - - -

6962 Woodstock Road
Quechee, Vermont
802.698.8675

- - - - -

www.singletonsvt.com

LEARN MORE ON FARMPLATE.COM

SOUTH ROYALTON MARKET

Beers + homebrew supplies in a food co-op

Located on the village green in the historic village of South Royalton, the market is a full-service grocery store with two beer-related departments. Their homebrew department is stocked with basic equipment and ingredients for every one from the beginner to the expert, regardless of style. Anything that's not on the shelves can be special-ordered. The beer and wine department has a broad representation of local breweries, including Long Trail, Wolaver's, Otter Creek, The Shed and Rock Art, as well as a great selection of beers from around the world. They also sell hard ciders and unusual, slightly eclectic wines.

Fun Fact: In addition to the department manager, several staff members are homebrewers. There's usually someone around who can answer your beer-related questions.

Don't Forget: The market carries local, conventional and organic products in every department. They also offer homemade sandwiches, salads and meals to go.

222 Chelsea Street
South Royalton, Vermont

- - - - -

802.763.2400

- - - - -

www.soromarket.com

ST. J. FOOD CO-OP

Eat better

The St. J. Food Co-op is a community-based, co-operatively owned natural foods store. It is a great place to buy fresh, local and organic foods, gather to meet and make friends and be part of a place that aspires to meet the needs of their greater community. Their collection of craft beers highlights the seasonality of Vermont microbreweries and they represent breweries from around the state—from their neighbors at Trout River Brewing *(see page 131)* to the organic ales brewed by Wolaver's *(see page 111)* in the Champlain Valley.

Fun Fact: The St. J. Food Co-op opened in 1998 after four years of outreach to garner community support.

Don't Forget: The co-op is happy to place a special order for almost any product.

490 Portland Street
St. Johnsbury, Vermont

- - - - -

802.748.9498

- - - - -

www.stjfoodcoop.com

STEEPLE MARKET

Food for every day to gourmet

Steeple Market is a unique grocery store located in the center of Fairfax, right in the heart of the community. They provide fresh, quality hand-cut meats and produce as well as a wide selection of groceries, specialty and local products, beer and wine. The market also has a full-service kitchen that spins out gourmet hot meals to go, pizzas, subs and more. Step inside the massive walk-in beer cave for one of the largest selections of craft beer in the area. They also have a growler filling station with three taps focused primarily on local Vermont brews with an occasional import.

Fun Fact: Steeple Market was the first store in the area to put in a growler filling station!

Don't Forget: The market offers delicious, made-to-order dinner specials Monday through Thursday that you can take home and serve!

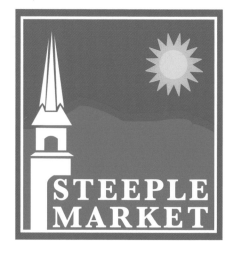

1098 Main Street
Fairfax, Vermont

- - - - -

802.849.6872

- - - - -

www.steeplemarket.com

STOWE BEVERAGE

Hundreds of craft brews right on the Mountain Road

Located on the Mountain Road in Stowe (the drinking town with the skiing problem), Stowe Beverage has been owned by the Mugford family for more than 20 years. They have an extensive selection of craft beers and specialize in Vermont beers, including The Alchemist's Heady Topper, Grassroots, Lost Nation, Rock Art, Otter Creek, Wolaver's, Long Trail, Magic Hat, Trout River, Harpoon and Switchback. They also have a great selection of craft brews from across the nation and around the world, including Logsdon Farmhouse Ales from Oregon, Prairie Artisan Ales from Oklahoma and from Europe, Mikkeller, Fantôme, Drie Fonteinen and occasionally Cantillon.

Fun Fact: Beers of the old world haven't been forgotten—Stowe Beverage carries more than 40 Belgian ales, dozens of German lagers and many beers from the UK, Italy, Spain and the Czech Republic.

Don't Forget: Only Visa and MasterCard (and of course cash and local checks) are accepted!

STOWE BEVERAGE & LIQUOR STORE

LIQUOR • BEER • WINE

Tel. 253-4525
1880 Mountain Road, Stowe. Open 9-9 M-S • 11-6 Sunday

1880 Mountain Road
Stowe, Vermont

- - - - -

802.253.4525

- - - - -

www.facebook.com/StoweBeverage

LEARN MORE ON FARMPLATE.COM

TERRILL STREET DISCOUNT BEVERAGE

Beer + wine, that's what we do!

Terrill Street Discount Beverage is Rutland County's premier, full-service beverage store. This family-owned business has the area's largest selection of domestic and imported wines as well as a full range of local, national and imported craft beers. In addition to their selection of bottled beers, they offer six beers on tap for growler fills. The store also has a curated section with the family's favorite Italian pastas, sauces and cheeses.

Fun Fact: Co-owner Bob Colloutti makes his own beer cheese spread from a recipe developed by his mother. He has quite the local following, so catch some if you can!

Don't Forget: The store often has "blowout" beer sales. Stop by frequently, sign up for their email list or watch for announcements on Facebook and Twitter so you're sure not to miss these great beer deals!

11 Terrill Street
Rutland, Vermont

- - - - -

802.775.1652

- - - - -

www.terrillstreetbeverage.com

U SAVE BEVERAGE

Deli + filling station

U Save Beverage has been family owned since 1995. For the last six years, their focus has been on Vermont's rapid growth in the craft beer industry, and they cater to the beer connoisseur in us all. They offer a draft system for growler pours called The Filling Station from which they sell small-batch beers that are hard to get unless traveling around the state. Usually three of the five taps feature Vermont beers and ciders. They are centrally located in the greater Burlington area near the airport, four major colleges and five microbreweries.

Fun Fact: At any given time, U Save carries more than 150 different craft beers from across the country and around the world.

Don't Forget: Monday through Friday you can grab one of their daily lunch specials while picking up a growler!

1332 Williston Road
South Burlington, Vermont

- - - - -

802.862.2907

THE VERMONT COUNTRY STORE

Purveyors of the practical + hard to find

- -

Visit the cheese shop at each of the two retail locations to discover unique Vermont craft brews that change with the seasons. In addition to a wonderful selection of Vermont specialty foods and drinks, The Vermont Country Store is stocked to the rafters with an enchanting variety of practical and hard-to-find goods. Today, the tradition of this family-owned business is carried on by Lyman Orton and his sons Cabot, Gardner and Eliot, seventh and eighth generation Vermonters and fourth and fifth generation storekeepers.

- -

Fun Fact: In 1945, when Vrest and Mildred Orton mailed their first catalog to the 1,000 people on their Christmas card list, they had no idea their modest offering of "36 Items You Can Buy Now" would grow into an enduring American company loved by customers around the world.

- -

Don't Forget: The Bryant House Restaurant at the Weston store also serves Vermont craft beer, both bottled and on tap.

657 Main Street (Route 100)
Weston, Vermont
802.824.3184

- - - - -

1292 Rockingham Road (Route 103)
Rockingham, Vermont
802.463.2224

- - - - -

www.vermontcountrystore.com

VILLAGE GROCERY

Where the grass is always greener

The Village Grocery began operating in 1939 as a full-service gas station. Today, they are trying to bring service back into the convenience store model, and they do their best to listen to customers. Their goal is to give back to the community that they, as a company, rely on— they focus on their local customers and they strive to bring in products from their friends and neighbors. Local beers and wines are a big part of their business, and they are always looking for creative ways to expand their offerings.

Fun Fact: The Village Grocery often has a hidden selection of local beer that is reserved for locals as well as tourists who are willing to share great stories!

Don't Forget: Every Friday before Columbus Day, someone from the Village Grocery drives to Maine to pick up fresh lobsters for their annual lobsterfest. So mark your calendars!

VILLAGE GROCERY
WAITSFIELD, VERMONT

4348 Main Street
Waitsfield, Vermont

- - - - -

802.496.4477

- - - - -

www.villagegrocery.info

WAITSFIELD WINE SHOPPE

Mad River Valley's purveyor of fine wine + craft beer

- -

The Waitsfield Wine Shoppe was created by Joan Wilson to help wine connoisseurs and enthusiasts alike find and taste hard-to-find, small-production wines at competitive prices. Each bottle represents the very best of its kind in terms of quality and value. The shop also offers more than 150 craft beers from around the world, including local Vermont favorites.

- -

Fun Fact: The shop is housed in the old Waitsfield post office building on Main Street.

- -

Don't Forget: Waitsfield Wine Shoppe has one of the most extensive wine selections in Vermont.

4330 Main Street
Waitsfield, Vermont
- - - - -
802.583.9463
- - - - -
www.waitsfieldwine.com

THE WARREN STORE

When visiting the Mad River Valley, a stop at The Warren Store is not to be missed. In true general store fashion, the front room has everything from penny candy and Vermont specialty foods to a diverse curated wine selection. At the back is an open kitchen where deli chefs prepare delicious meals from scratch, with an emphasis on locally grown and produced ingredients. Enjoy breakfast or lunch inside or on the deck overlooking the river. You can also pick up dinner to go. The store has a growing selection of "ARTisanal" beer and hard cider, featuring Lawson's Finest among many others.

Fun Fact: The store is renowned for organizing one of the most fun, funky Fourth of July celebrations in the state.

Don't Forget: The Warren Store hosts a Heroes of Hops event at the Pitcher Inn each August to celebrate all the great Vermont brewers.

The local store and so much more.

284 Main Street
Warren, Vermont

- - - - -

802.496.3864

- - - - -

www.warrenstore.com

LEARN MORE ON FARMPLATE.COM

257

WILLEY'S STORE

Serving NEK communities for over 100 years

Willey's Store is one of Vermont's largest and oldest country stores. It has been family owned and operated for five generations. You will find everything you need under one roof—grocery items, hardware, clothing and dry goods—and there's even a gas pump outside. They are proud to carry a diverse collection of local beer and other local products, including yarn and grass-fed beef and pork.

Fun Fact: Every year Willey's ships hundreds of flour-sack towels to customers all over the United States.

Don't Forget: You can stay up-to-date with all the news and goings on via their Facebook page.

7 Breezy Avenue
Greensboro, Vermont

- - - - -

802.533.2621

- - - - -

www.facebook.com/willeys.store

WILLOUGHBY LAKE STORE

A traditional country store serving Lake Willoughby

Willoughby Lake Store, formerly known as The Millbrook Store, has been taking care of the needs of locals, leaf peepers, fishermen, hunters and summer visitors for well over 100 years. They offer a variety of groceries, hardware, clothing, antiques and collectibles. The deli serves hot and cold sandwiches, soups and meat and cheese by the pound. The kitchen stays hot cooking homemade pizzas, fresh baked muffins, cookies and pastries. In the summer, more than 20 flavors of hard and soft ice cream fly out the ice cream window.

Fun Fact: Beer has become an important part of the Willoughby Lake Store as owners Jim and Sandy research and buy some of the best beer brewed. Customers are often surprised to see such a wide selection of beer. Vermont has certainly earned its place as brewing some of the best beers in the world, but the store's full selection of beer does not stop there—they offer beers from around the world.

Don't Forget: Their wine collection is growing as well. They offer local Vermont wines as well as imports from as far as Australia and New Zealand.

2003 Route 5A
Orleans, Vermont
- - - - -
802.525.3300

WOODSTOCK FARMERS' MARKET

Eatin' fresh, eatin' local

The Woodstock Farmers' Market is a busy, year-round fresh market filled with delicious food and drink. They specialize in farm-fresh produce, artisanal products, creative sandwiches and prepared foods along with carefully selected craft beers and wines. Their beer mantra is to support craft breweries "from nearby," and they often know the people who brew the beer they carry! They also have a selection of exceptional beers from around the world. WFM believes that artisan beer producers are critical to the success of the local food movement as they help raise the beer/food standard. Their mission is to keep that connection alive for the community, whether the beer is brewed in Bridgewater, Vermont, or in Yorkshire, England.

Fun Fact: The market has produced two beers under the WFM brand in partnership with Rock Art Brewery *(see page 115)* of Morrisville, Vermont—a craft brewer of fine ales. Their first beer was a brown ale called WFM 20th Anniversary Farmers Brown Ale; their second one is called Farmer's Simcoe and is a single varietal hopped ale.

Don't Forget: Check out their cool selection of unique 22-ounce and other large-format beers. They have more than 20 varieties representing several countries.

979 West Woodstock Road
Woodstock, Vermont

- - - - -

802.457.3658

- - - - -

www.woodstockfarmersmarket.com

WOODSTOCK HOPS N' BARLEY

A craft beer + wine shop

Woodstock Hops N' Barley is the Upper Valley's premier craft beer retail destination. They specialize in bringing the most interesting craft beers available in the state to their customers—offering everything from super-fresh session IPAs to funky saisons and barrel-aged barley wines and stouts. They have worked extremely hard with local distributors to ensure they have access to the most elusive brews. In addition, they have a small but unique handpicked selection of wines. Their creative wine rack is organized from light and crisp to rich and full bodied, much like an actual wine-tasting setup. Stop in to chat with their knowledgeable staff—they can help you find just what you are looking for.

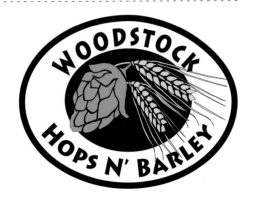

Fun Fact: Woodstock Hops N' Barley features 10 growler lines, making it one of the top retail destinations in the state in terms of the number of brews on tap.

Don't Forget: They are only able to fill glassware with their own logo. It is available for a $5 deposit.

446 Woodstock Road
Woodstock, Vermont

- - - - -

802.457.2472

- - - - -

www.woodstockhopsnbarley.com

EXTRAS

ON TAP SOON

Bent Hill Brewery
Braintree, Vermont
www.facebook.com/benthillbrewery

Brocklebank Craft Brewing
Tunbridge, Vermont
www.brocklebankvt.com

Burlington Beer Company
Burlington, Vermont
www.burlingtonbeercompany.com

Canteen Brewery
Waitsfield, Vermont
www.facebook.com/canteenbrewing

Cousins Brewing Company
Waitsfield, Vermont
www.localfolkvt.com

Freight House Brewing
South Royalton, Vermont
www.facebook.com/freighthousebrewing

Hop n' Moose Brewing Company
Rutland, Vermont
www.facebook.com/hopn.moose

Knotty Shamrock
Northfield, Vermont
www.knottyshamrock.com

Infinity Brewing
South Burlington, Vermont
www.infinitybrewing.com

Prohibition Pig
Waterbury, Vermont
www.prohibitionpig.com

Queen City Brewery
Burlington, Vermont
www.queencitybrewery.com

BEER EVENTS

SEPTEMBER

Croptoberfest
Stowe, Vermont
www.cropvt.com

Mount Snow Brewers Festival
West Dover, Vermont
www.mountsnow.com/event

Siptemberfest
Fayston, Vermont
www.siptemberfest.com

Trapp Family Lodge Oktoberfest
Stowe, Vermont
www.trappfamily.com

OCTOBER

Green Mountain Octoberfest
Hancock, Vermont
**www.facebook.com/
greenmountainoctoberfest**

Harpoon Octoberfest Vermont
Windsor, Vermont
www.harpoonbrewery.com/events

Bean & Brew Festival
Jay, Vermont
www.jaypeakresort.com/events

Killington Brewfest Weekend
Killington, Vermont
www.killington.com/events

Magic Hat Oktoberfeast
South Burlington, Vermont
www.magichat.net/oktoberfeast

Oktoberfest at Mount Snow
West Dover, Vermont
www.mountsnow.com/event

TOURS + TRAILS

Burlington Brew Tours
802.760.6091
www.burlingtonbrewtours.com

Burlington Hikers & Beer Lovers
www.meetup.com/burlington-hikers-and-beer-lovers

The Good Bus
802.776.8333
www.thegoodbus.com

Sojourn's Vermont Bike and Brew
800.730.4771
www.gosojourn.com/vermont-bike-brew-tour

Vermont Backroad Brewery Tours
802.446.3131
www.vtbackroadtours.com/brewery.html

Dig In Vermont Trails
www.diginvt.com/trails

Vermont Bike & Brewery Adventure
888.448.5876
www.mountmajor.com/vermont-bike-and-brewery

Vermont Brewery Challenge
802.885.1262
www.vermontbrewers.com/passport-program

HOMEBREWING SUPPLIES

The Beverage Den & Smokeshop
340 North Street
Bennington, Vermont
802.442.2861

Brewfest Beverage
199 Main Street
Ludlow, Vermont
802.228.4261
www.brewfestbeverage.com

Bristol Discount Beverage
21 Prince Lane
Bristol, Vermont
802.453.3990

Craft Beer Cellar of Waterbury
3 Elm Street
Waterbury, Vermont
802.882.8034
www.craftbeercellar.com

Joe's Discount Beverage
335 River Street
Springfield, Vermont
802.885.3555

Local Potion Homebrew Suppliers
20 School Street
Plainfield, Vermont
802.454.7200
www.localpotion.net

South Royalton Market
222 Chelsea Street
South Royalton, Vermont
802.763.2400
www.soromarket.com

Vermont Homebrew Supply
147 East Allen Street
Winooski, Vermont
802.655.2070
www.vermonthomebrew.com

Woodstock Hops N' Barley
446 Woodstock Road
Woodstock, Vermont
802.457.2472
www.woodstockhopsnbarley.com

LOCAL HOPS + GRAINS

Addison Hop Farm
Addison, Vermont
802.989.4214

Anjali Farm
South Londonderry, Vermont
www.anjalifarm.com

Border View Farm
Alburgh, Vermont
802.796.3292

Butterworks Farm
Westfield, Vermont
www.butterworksfarm.com

Hoppy Valley Organics
Pownal, Vermont
www.facebook.com/hoppyvalleyorganics

Northeast Hop Alliance
www.northeasthopalliance.org

Square Nail Hops Farm
Ferrisburgh, Vermont
802.355.3309

Sunnybrook Farm
Middlesex, Vermont
sunnybrookhops@gmail.com

The Vermont Hops Project
Burlington, Vermont
www.uvm.edu/extension/cropsoil/hops

Walden Heights Nursery & Orchard
Walden, Vermont
www.waldenheightsnursery.com

RESOURCES

American Brewers Guild
www.abgbrew.com

American Homebrewers Association
www.homebrewersassociation.org

Brewers Association
www.brewersassociation.org

Burlington Beer Enthusiasts
www.meetup.com/burlington-beer-enthusiasts

The Green Mountain Mashers
Homebrew Club of Vermont
www.mashers.org

Homebrewer Outreach &
Preservation Society (HOPS)
www.hopsvt.org

Vermont Beer
www.vtbeer.org

Vermont Brewers Association
www.vermontbrewers.com

Homebrew Guru of Vermont
www.vthomebrewguru.com

TASTING NOTES